Everything you always wanted to know about...

Environmental Science

2nd edition

STERLING
Education

Our guarantee – the highest quality educational books.

Be the first to report a typo or error and receive a
$10 reward for a content error or
$5 reward for a typo or grammatical mistake.

info@sterling–prep.com

We reply to all emails – please check your spam folder

2   1

ISBN-13: 978-1-9475566-4-5

Sterling Education
6 Liberty Square #11
Boston, MA 02109

info@sterling-prep.com

Published by Sterling Education

 Printed in the U.S.A.

Dear Reader!

From the foundations of Earth systems to the present-day climate challenges, this clearly explained text is a perfect guide for anyone who wants to be knowledgeable about environmental science. This book is aimed at providing readers with the information necessary to make them more engaged and appreciative participants in the global environment.

This book was designed for those who want to develop a better understanding of ecosystems, population dynamics, use of natural resources, as well as the political and social landscape of environmental challenges. The content is focused on an essential review of all the important facts and events shaping the natural world we live in.

You will learn about Earth's biochemical cycles, land and water use, energy resources and their consumption, the significance of the various environmental movements and global initiatives, as well as how different human actions affect the overall balance within ecosystems.

Created by highly qualified science teachers, researchers, and education specialists, this book educates and empowers both the average and the highly informed readers, helping them develop and increase their understanding of environmental problems and solutions.

We congratulate you on your desire to learn more about environmental science. The editors sincerely hope that this guide will be a valuable resource for your learning.

vfd191118

Our Commitment to the Environment

Sterling Test Prep is committed to protecting our planet's resources by supporting environmental organizations with proven track records of conservation, ecological research and education and preservation of vital natural resources. A portion of our profits is donated to help these organizations so they can continue their critical missions. These organizations include:

 For over 40 years, Ocean Conservancy has been advocating for a healthy ocean by supporting sustainable solutions based on science and cleanup efforts. Among many environmental achievements, Ocean Conservancy laid the groundwork for an international moratorium on commercial whaling, played an instrumental role in protecting fur seals from overhunting and banning the international trade of sea turtles. The organization created national marine sanctuaries and served as the lead non-governmental organization in the designation of 10 of the 13 marine sanctuaries.

 For 25 years, Rainforest Trust has been saving critical lands for conservation through land purchases and protected area designations. Rainforest Trust has played a central role in the creation of 73 new protected areas in 17 countries, including the Falkland Islands, Costa Rica and Peru. Nearly 8 million acres have been saved thanks to Rainforest Trust's support of in-country partners across Latin America, with over 500,000 acres of critical lands purchased outright for reserves.

 Since 1980, Pacific Whale Foundation has been saving whales from extinction and protecting our oceans through science and advocacy. As an international organization, with ongoing research projects in Hawaii, Australia, and Ecuador, PWF is an active participant in global efforts to address threats to whales and other marine life. A pioneer in non-invasive whale research, PWF was an early leader in educating the public, from a scientific perspective, about whales and the need for ocean conservation.

With your purchase, you support environmental causes around the world.

Table of Contents

Table of Contents (*continued*)

Chapter 1

Earth Systems and Resources

Earth Science Concepts

- Geologic time scale

- Plate tectonics

- Earthquakes

- Volcanism

- Seasons

- Solar intensity and latitude

The Atmosphere

- Composition

- Structure

- Weather and climate

- Atmospheric circulation and the Coriolis Effect

- Atmosphere-ocean interactions

- ENSO

Global Water Resources and Use

- Freshwater/saltwater

- Ocean circulation

- Agricultural, industrial and domestic use

- Surface and groundwater issues

- Global problems

- Conservation

Earth Science Concepts

Earth science attracts many people because they love the outdoors, and many scientists want to study the processes that create and modify the landforms of Earth. Some may want to research more in-depth what drives the surface processes and other features of the planet; for example, why do earthquakes occur, or volcanoes erupt? These scientists are interested in understanding the layers of material that lie beneath the surface, the mantle and the core, and the processes going on below the surface.

Oceans cover more than 70% of the Earth; therefore, it is not surprising that many ponder about what lies within the oceans and at the bottom of the sea. Some scientists say that more is known about the far side of the Moon than about the deep oceans; however, a lot is known considering how hostile the ocean environment can be for land-dwelling species.

Most earth scientists specialize in studying only one facet of the planet, as the Earth is a very large, complex system and subsets of systems. Scientists from all branches of earth science work together to answer narrowly defined questions since all branches of earth science are interconnected. The following sections will discuss the different branches and concepts of earth sciences.

Geologic time scale

The origins of Earth's geologic time scale

Although mining has been of commercial interest since the time of ancient civilizations, it was not until the 1500s and 1600s that mining became an industry to generate income. These men who worked the mines were perhaps some of the first people to acquire an understating of the geological relationships of different rock types.

In 1669, Nicolaus Steno described two basic geologic principles by noting the relationships of different rock layers. Steno's first principle stated that sedimentary rocks are laid horizontally; his second principle was that younger rock layers were deposited on top of older rock layers. Hundreds of new layers have formed over the past several million years, with sediment settling and being compressed to form a new section on top of the old layer. This occurs in both drylands and at the bottom of the oceans.

In 1795, James Hutton introduced an additional concept called the *principle of uniformitarianism,* which was the idea that natural geologic processes were uniform in frequency and magnitude throughout time. These principles allowed mine workers and scientists of the time to start recognizing distinct types of rock successions within the layers of the Earth. However, comparisons between rock sequences of different areas were often not possible because they categorized rocks by their color, texture or even smell. They were able to use fossils to correlate between geographically distinct areas because fossils can be found over wide regions of the Earth's crust.

Another major contribution to the geologic time scale came from a surveyor, canal builder, and amateur geologist from England named William Smith. In 1815 he produced a geologic map of England and validated the legitimacy of the principle of "faunal succession." This principle states that fossils are found in rocks in a precise order. The "faunal succession" principle led other scientists to use fossils to define large increments within a relative timescale.

Dividing the Earth's History into Time Intervals

Geologists have divided the geologic history of Earth into a series of time intervals. However, these time intervals are not equal in length, such as the hours in one Earth day. Instead, the time intervals are variable in length because scientists divide geologic time using significant events in the history of the Earth, called "boundary events."

Examples of boundary events

One example, the Marine Realm, and the End-Permian Extinction marked the end of the Permian period. It was notable for the greatest mass extinction of the last 600 million years, during which possibly 90% of marine animal species and 70% of land species disappeared. The theoretical causes of this mass extinction range from the creation of the Pangaea supercontinent to volcanism.

Another example of a boundary event was the Cretaceous–Paleogene Extinction event that happened around 66 million years ago. It was a mass extinction of around three-quarters of all plant and animal species on Earth, which included all non-avian dinosaurs and occurred over a geologically short period. This mass extinction marked the end of the Cretaceous period and, with it, the entire Mesozoic Era, starting the Cenozoic Era that continues to this day.

The Earth's geologic time scale divisions

The geologic time scale of the Earth is broken up into categorized segments of time. From the largest to smallest, this order includes eons, eras, periods, epochs, and ages. These are indicated in the top portion of the geologic time scale shown below.

Table 1. Divisions of the Earth's geologic time scale.

Eon	Era	Period	Epoch	Age
				Chattian
Phanerozoic	Cenozoic	Paleogene	Oligocene	Rupelian

Eons are the largest intervals of geologic time and last for hundreds of millions of years in duration. As seen in the timescale above, the Phanerozoic Eon is the most recent eon and began more than 500 million years ago.

Eras are eons that are divided into smaller time intervals. The Phanerozoic period is divided into three eras: Cenozoic, Mesozoic, and Paleozoic. Scientists use very significant events in the Earth's history to determine the boundaries of each era.

Periods are eras that are subdivided. Some events bind the periods over a wide extent, but they are not as significant as those that bind the eras. As can be seen in the time scale below, the Paleozoic era is subdivided into the Permian, Pennsylvanian, Mississippian, Devonian, Silurian, Ordovician, and Cambrian periods.

Epochs are finer subdivisions of periods of an era. The subdivision of periods into epochs can be done only for the most recent part of the geologic time scale because the older rocks are deeply buried, intensely deformed, and severely modified by long-term Earth processes.

The geologic history contained within the older rocks cannot be interpreted as clearly.

GSA Geologic Time Scale

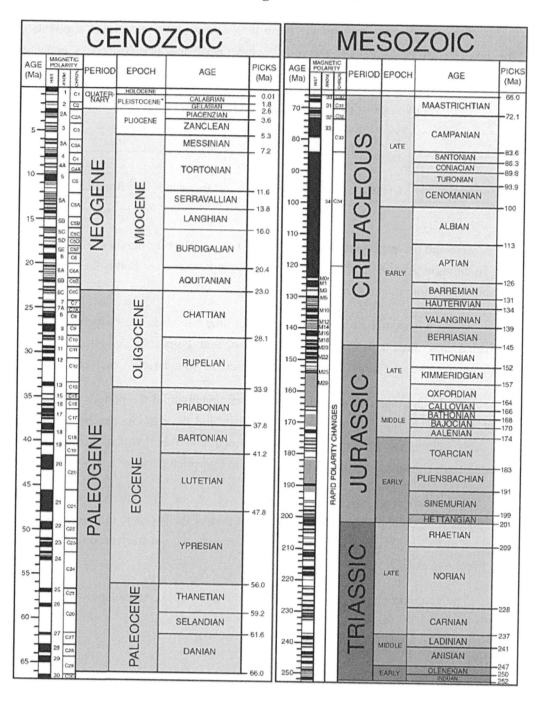

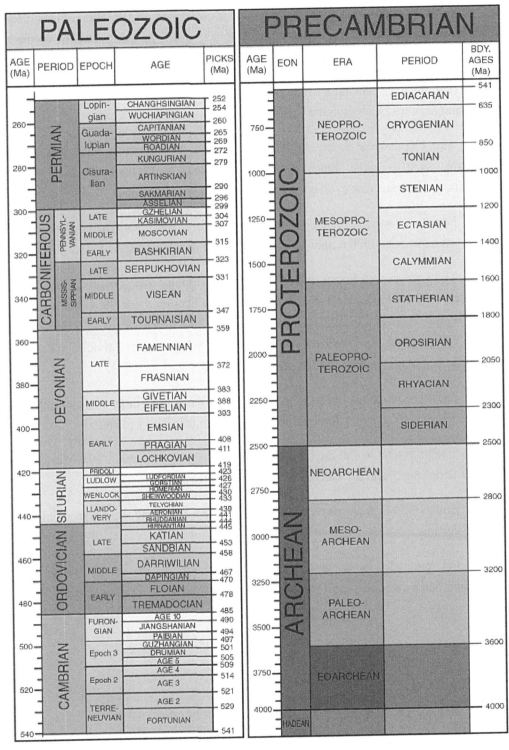

Figure 1.1 (a, b). *A geological time scale constructed by The Geological Society of America.*

Scientists constructed the geologic time scale to show the duration of each period in the history of the Earth. They accomplished this by making a linear timeline on the left side of the time columns, shown in the scheme below.

The thicker sections, such as the Proterozoic Eon, were longer in duration than the thinner sections, such as in the Cenozoic Era.

The Phanerozoic Eon represents the period during which most macroscopic organisms, such as algae, fungi, plants, and animals lived, and coincides with the appearance of animals that evolved external skeletons (exoskeletons).

Scientists usually refer to the time before the Phanerozoic Eon as the Precambrian Period, which is shown on the right side of the time scale below.

The Phanerozoic Eon is subdivided into three major divisions: the Cenozoic, the Mesozoic, and the Paleozoic Eras. The suffix "zoic" comes from the Greek root word "zōē," which means "life." "Cen-" is the Greek root word "καινός," which means "recent." "Meso-" is the Greek root word "mésos," which means "middle." "Paleo-" is the Greek root word "palaio," which means "ancient."

The three divisions reflect major changes in the composition of ancient life, with each era characterized by its domination by a group of animals or plants.

Scientists tend to call the Cenozoic Era the "Age of Mammals," the Mesozoic the "Age of Dinosaurs," and the Paleozoic the "Age of Fishes."

However, this is an overly simplified view that has some significance but can be a bit misleading.

For example, other groups of animals, such as mammals, turtles, frogs, crocodiles, and a vast number of insects, lived on land during the Mesozoic Era.

Plate tectonics

Three types of plate boundaries occur on Earth. In addition to the three types of boundaries, there are *plate boundary zones* that are broad belts in which boundaries are not well defined, and the effects of plate interaction are unclear.

- *Divergent boundaries* occur when two tectonic plates move away from each other, and the movement of the plates pulling away from each other creates a new oceanic crust. This crust is made of basalt.

- *Convergent boundaries* occur when one plate dives under another, destroying the crust.

- *Transform boundaries* occur where the crust is neither produced nor destroyed as the plates slide horizontally past each other.

When two tectonic plates move away from each other, a *divergent boundary* forms. First lava spews from long fissures on the ocean floor. While this happens, geysers are spurting superheated water alongside these boundaries, and frequent earthquakes occur along the boundary rift. Magma or molten rock then rises from the mantle just beneath the rift.

Magma exudes up into the gap and hardens into solid rock, which forms a new crust on the torn edges of the tectonic plates. Magma coming up from the mantle solidifies into basalt – a dark, dense rock that lies beneath the ocean floor. Consequently, oceanic crust (which is made of basalt) is created at divergent boundaries.

A *convergent boundary* occurs when two plates come together. The impact of the two tectonic plates causes the edge of one or both tectonic plates to buckle. The buckled plates then push one or both edges up, forming a rugged mountain range. One buckled plate can sometimes force the other plate down deep into a seafloor trench. Often a chain of volcanoes forms parallel to the boundary, the mountain range, or the trench. When this occurs, powerful earthquakes shake a wide area on both sides of the boundary.

Sometimes, one of the colliding tectonic plates is topped with oceanic crust. When this happens, the tectonic plate is forced down into the mantle and starts to melt. Then, magma rises into and through the other plate, hardening into a new crust.

The magma that formed from the melting tectonic plates hardens into granite — a light-colored, low-density rock — and makes up the continents of the Earth. Therefore, continental crust (made of granite) is created, and oceanic crust is destroyed at convergent boundaries.

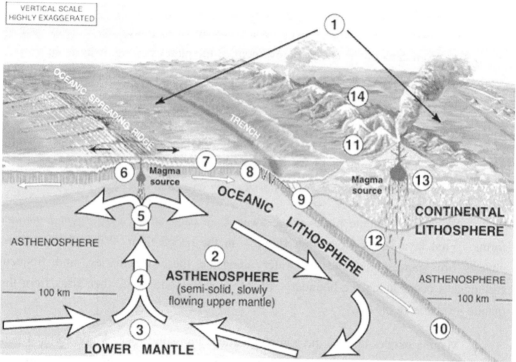

Figure 1. 2. *Convergent (11) and divergent (5) plate boundaries that occur in oceans and land.*

When two tectonic plates slide past each other, the formation is called a *transform plate boundary*. This type of boundary will split and carry away any natural or human-made structures in the opposite direction.

As the plates grind against each other, rocks that line the boundary are crushed, which creates a linear fault valley or undersea canyon. Earthquakes rattle through a wide boundary zone.

In comparison to convergent and divergent boundaries, the magma does not form. Consequently, the crust is not created nor destroyed but cracked and broken at transform margins.

Earthquakes

Terminology:

Fault plane, or fault, is the area of the tectonic plate that slips, causing an earthquake.

Hypocenter is the area below the surface of the Earth where an earthquake originates.

Epicenter is the point of origin of an earthquake on the Earth's surface (directly above the hypocenter).

Foreshocks are smaller earthquakes that occur in the same area before the large earthquake takes place. Scientists are currently unable to predict if foreshocks are the precursors to a larger earthquake or just small, independent quakes.

Mainshocks are large earthquake events that often cause the most damage.

Aftershocks are smaller earthquakes that occur after a mainshock. If the mainshock is very large, the aftershocks may continue for days or even years.

The Earth is composed of four main layers: the inner core, the outer core, the mantle, and the crust. The mantle and the crust make up a relatively thin layer known as the surface. They occur in "broken" parts that cover much of the Earth's surface, much like puzzle pieces that roughly fit together. These pieces are termed *tectonic plates*, and their edges are known as *plate boundaries*.

These plate boundaries are made up of many faults, which are the prime location for earthquake activity. Their edges are very rough; they often get caught on the boundary of another plate. While the sides or edges of the faults are caught on one another, the rest of the block continues to move. The energy that would normally be expended in moving the plates becomes stored up as potential energy in the areas where the plates are caught.

When the force of the plates becomes large enough to overcome the friction that caused them to stop, the edges unstick, and the potential energy is released. This energy radiates outward from the hypocenter in all directions as *seismic waves*, much like ripples in a pond radiate outwards when a fish breaks the surface. These waves shake the Earth's surface as they move through the crust, causing damage both above and below the surface.

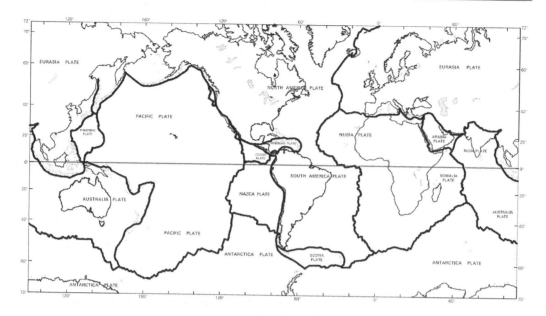

Figure 1.3. *A simplified map of the tectonic plates that span the Earth.*

Earthquakes are recorded via *seismographs*. The base rests on the ground while a heavy weight hangs free. When the Earth shakes, the base of the seismograph and the coil or string holding the weight absorbs all the movement. A *seismogram* is a recording that is made by a pen that is attached to the underside of the weight.

The actual size of an earthquake depends on both the size of the fault and the amount that it slips. Since this distance cannot be simply recorded with a measuring tape, a seismogram is used to measure its size, or *magnitude*. This is represented by a number on the Richter Scale, a base-10 logarithmic scale that was created by Charles F. Richter in 1935 to compare the size of earthquakes.

The magnitude of an earthquake is determined by taking the logarithm of the amplitude of waves (recorded by seismographs), and it is expressed in whole numbers and decimals.

Since earthquakes are measured on a logarithmic scale, each whole-number increase in magnitude (e.g., from 5.3 to 6.3) represents a tenfold increase in amplitude.

Scientists estimate that a jump in one whole integer on the Richter Scale will result in the release of nearly 31 times the amount of energy.

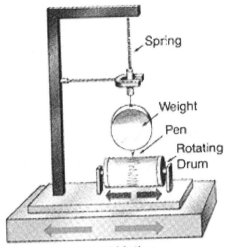

Horizontal Motion

Figure 1.4. *The base of a seismograph and the spring will rock with the motion of the Earth, as the weight stays still. The pen on the bottom side of the weight will record the displacement of the base with respect to the weight.*

Seismograms are also used for locating the origin of earthquakes. Scientists use a method called triangulation to do this, which requires three seismograph readings.

A circle is drawn on a map around the station that houses a seismograph, with the radius of each circle the distance from the station to the area where the earthquake occurred.

The location on the map where the three circles intersect is the epicenter of the earthquake.

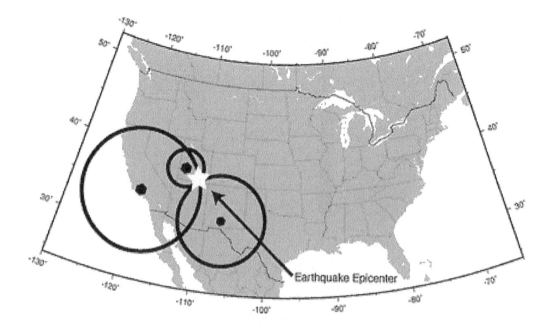

Figure 1.5. *Triangulation method to determine the epicenter of an earthquake.*

Though quakes may happen more frequently along certain faults, scientists are unable to predict when or where an earthquake will occur.

Some have claimed that certain weather patterns may exist or that animals can sense a change before an earthquake takes place.

However, these events are not consistent and are given little credibility by the scientific community.

Volcanism

Terminology:

Subduction is a process that occurs at a convergent plate boundary when one tectonic plate moves under another and plunges into the mantle.

Orogenesis, often called mountain building, takes place when large pieces of Earth sitting on top of the subducting plate are pressed into the plate on top.

Volcanism is defined as the eruption of lava and other material from deep within the Earth. Over time, the eruptions of material form a mountainous structure called a volcano. Though volcanoes may be the most commonly discussed or well-known type of volcanism, most volcanic eruptions do not form volcanoes.

Volcanism recycles material that travels through plate tectonics. Everything that is subducted below the crust is believed to return to the surface as molten magma. When magma erupts and becomes lava, it returns gases, fluids, and solids to the Earth's surface.

These solids include *igneous rock*, which is now available to become sedimentary and metamorphic rocks. The *fluid* products of erupted magma release water as they rise and contain silica, metals, and other elements. This magmatic fluid can change the composition of rocks, ore deposits, and sulfate minerals. *Gaseous* elements, such as sulfur gases, water vapor, and carbon dioxide, are what cause the lava to erupt. They cause magma to expand, sending it to the surface quickly. Volcanic gases then enter the atmosphere and exert an influence on the overall atmospheric composition and the global climate.

Four Major Types of Magmatic Volcanism

Most volcanic eruptions occur underwater along mid-ocean ridges called *divergent margins*. The crust of the ocean is pulled apart via plate divergence, causing the hot rock in the mantle layer below to melt due to the decreased pressure around it. The part of the rock that melts is the *magma*, which is composed of basalt. What remains is a part of the mantle that is a heavier rock composed mainly of olivine, which scientists call *peridotite*. This most common type of volcanism occurs as a quiet oozing of basalt lava from the cracks in the ocean floor. There are a few cases of divergent volcanism on land, though they appear and act much differently than divergent oceanic margins.

There is a second type of volcanism that occurs along subduction zones (convergent margins) where oceanic plates, saturated with sediment and water, plunge into the hot mantle layer. This type of volcanism produces most of the world's volcanoes, called *composite cone volcanoes* or "stratovolcanoes." These are cone-shaped volcanoes made of hundreds of layers of lava, rock, ash, and debris. In this case, magma is not created by releasing pressure on the mantle but instead by adding water to it.

Subduction-created magma then rises into the lower crust where it will collect over time and occasionally produce a violent eruption. Mount St. Helens in Washington State is an example of a composite cone volcano.

The third type of volcanism, which includes about 10% of all volcanoes on Earth, is known as *hotspots*. Divergent or convergent plate boundaries do not form these. Geologists once thought that hotspots arose from deep within the mantle, but they now have a new theory. Scientists currently believe that lithospheric plates are fracturing, causing magmas to form under the release of pressure, much like what is seen in divergent settings.

The fourth and final type of magmatic volcanism no longer occurs today and is called *flood volcanism*. In this instance, copious amounts of basalt lava poured out of breaks and fissures, covering thousands of square miles. This occurred both under the sea and on land. Extensive knowledge is not currently known about flood volcanism, but intense studies are underway.

Non-Magmatic Volcanism

Non-magmatic volcanism is not as common and does not involve magma. One such example is mud volcanism, which occurs both on land and in the sea. On land, hundreds of mud volcanoes occur in areas where hydrocarbons are readily abundant, such as Azerbaijan and Trinidad.

Under the sea, hundreds of mud volcanoes can be found near subduction trenches, where an important component called *serpentinite mud* is found.

Seasons

Terminology

Solstice occurs twice a year and is the moment when the Earth's tilt is at its maximum. There are two times a year when this occurs:

Winter solstice is the shortest day and longest night of the year in the Northern Hemisphere, and it usually takes place on December 21. On this day, the Earth's tilt away from the Sun is at its maximum.

Summer solstice is the longest day and shortest night of the year, and it is when the Earth's tilt towards the Sun is at its maximum. It usually takes place around June 21.

There are two ways that the Earth moves. It is in constant motion around the Sun, completing its orbit once every 365.25 days. It also spins along its polar axis, making a complete rotation once every 24 hours. When these two motions intersect, *equinoxes* occur.

The mass of the Earth is incredible, which produces a gyroscopic effect that keeps its poles pointing in the same direction (though a large earthquake may cause the Earth to wobble very slightly on its axis). To make sense of these equinoxes, scientists mark the position of the Sun in the sky relative to the rotation of the Earth's poles.

The most important line that scientists reference is the celestial equator, which divides the Earth into Northern and Southern Hemispheres. The Sun will cross the celestial equator twice a year, moving from the Southern to the Northern Hemisphere and vice versa. These two crossings mark the vernal and autumnal equinoxes. In September, during the autumnal equinox, the Sun moves from the Northern Hemisphere to the Southern Hemisphere and passes overhead at every point along the Earth's equator. During this time, the Sun will rise exactly in the east and set directly in the west, causing the lengths of day and night to be roughly equal.

During the vernal equinox in March, the Sun will appear to cross the celestial equator, moving from the Southern to the Northern Hemisphere. During this equinox, the refraction of the Sun's rays will cause it to appear slightly above the horizon when the "true" center of the Sun is already setting or rising. This causes the days to begin to lengthen.

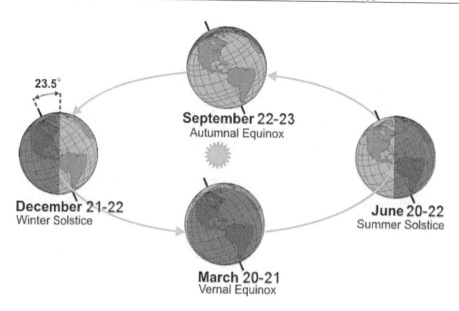

Figure 1.6. *Equinoxes and solstices and their time frames.*

A common misconception is that the Earth is closest to the Sun during the summer, causing the hot weather, and further from the Sun in the winter. Although this idea may seem sensible, it is not how seasons function. The Earth's orbit around the Sun is not perfectly circular; in fact, it looks very much like an oval.

During certain parts of the year, the Earth does get closer to and further away from the Sun. In the Northern Hemisphere, the Earth is the farthest away from the Sun during the summer months and the closest in the winter. Because of the extreme distance of the Earth from the Sun, heat from the Sun has little to do with the differences in seasons.

The reason for the changes in seasons on planet Earth has to do with the *axis*, an imaginary pole that runs from the "top" to the "bottom," or from the North Pole to the South Pole.

The Earth spins around this pole, making one complete spin in roughly 24 hours (hence the length of one day). Earth has seasons because its axis is not seated exactly vertically.

Earth has seasons because its axis is tilted. Earth rotates on its axis as it orbits the Sun, but the axis always points in the same direction

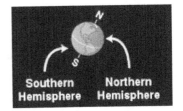

Southern Hemisphere   **Northern Hemisphere**

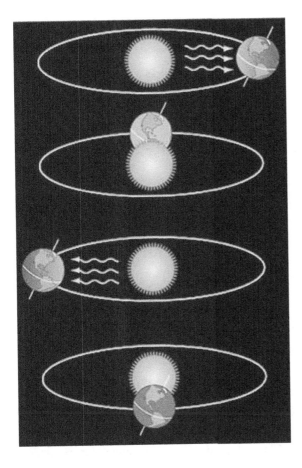

December:
Summer is south of the equator; winter is north of the equator. The Sun shines directly on the Southern Hemisphere and indirectly on the Northern Hemisphere.

March:
Fall is south of the equator; spring is north of the equator. The Sun shines equally on the Southern and Northern Hemispheres.

June:
Winter is south of the equator; summer is north of the equator. The Sun shines directly on the Northern Hemisphere and indirectly on the Southern Hemisphere.

September:
Spring is south of the equator; fall is north of the equator. The Sun shines equally on the Southern and Northern Hemispheres.

Figure 1.7. *Seasons are a product of the Earth's tilted axis and its path around the Sun.*

Solar intensity and latitude

Total Solar Irradiance is the maximum possible power that the Sun can deliver to the Earth at its average distance from the Sun (about 1,360 watts per square meter according to the most recent measurements by NASA scientists).

Since only half of the Earth is ever irradiated by the Sun at once, this cuts in half the amount of total solar irradiance.

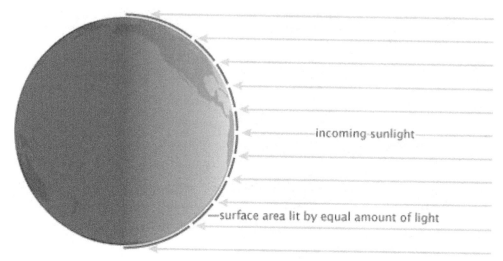

incoming sunlight

surface area lit by equal amount of light

Figure 1.8. *Total solar irradiance is the maximum power the Sun can deliver to a surface that is perpendicular to the path of the incoming rays of light.*

Since the Earth is a sphere, only the areas at the equator at midday come anywhere close to being perpendicular to the incoming light from the Sun. Every other place on Earth receives the Sun's rays at an angle. This progressive decrease in the solar illumination angle, along with the progressive increase of latitude, reduces the average solar irradiance by an additional 50%.

Incoming sunlight

All matter that has a temperature above absolute zero will radiate the energy of some wavelength that occurs in the electromagnetic spectrum. The hotter an object, the shorter the wavelength of its radiated energy.

The hottest objects in the universe, such as some stars, radiate mainly x-rays and gamma rays. Cooler objects will emit longer-wavelength radiation, including microwaves, thermal infrared waves, radio waves, and visible light.

The Sun's peak radiation is in the spectrum of visible light, and its temperature is over 5,500 °C. Over the entire planet, the average amount of sunlight that arrives at the top of the Earth's atmosphere is merely one-fourth of the total solar irradiance or about 340 watts per square meter.

If an incoming flow of solar energy is balanced by an equal flow of heat back into space, the Earth is said to be in radiative equilibrium, and the global temperature will remain relatively stable.

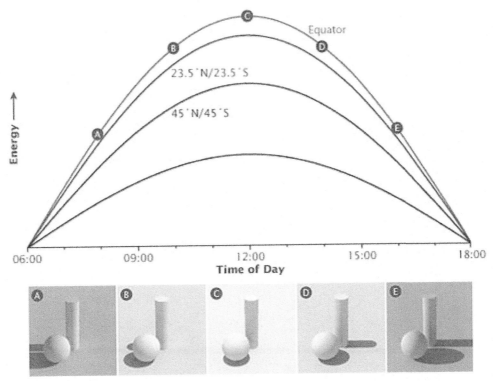

Figure 1.9. *The solar radiation received by the surface of the Earth varies depending on the time of day and the latitude at any given location. The above graph illustrates the relationships between time, latitude, and solar energy during the vernal and autumnal equinoxes.*

The pictures (A-E) show how the time of day affects the angle of incoming sunlight (shown by the length of the shadow) and the intensity of light.

During the two equinoxes, the Sun will rise at 6 am in all locations. The intensity of sunlight will increase from sunrise until noontime when the Sun is then directly over the equator and casting no shadows (C). After noon, the intensity of sunlight begins to decrease until the Sun sets at 6 pm.

Notes

The Atmosphere

The atmosphere that surrounds planet Earth is composed of many layers. Each layer has unique temperature changes, density, chemical composition, and movement.

All the gases in the atmosphere combine to absorb ultraviolet radiation that comes from the Sun, which can be very harmful to life on Earth.

The atmosphere also traps heat, which warms the planet's surface.

Terminology:

Rayleigh scattering is the phenomenon that causes the sky to look blue. As light moves into the atmosphere, most of the long wavelengths, such as orange, red, and yellow, pass through.

The short wavelengths of blue light get absorbed by molecules of gas and radiate in all directions. This scattered blue light can be seen from the surface, making the sky appear blue.

The Karman line is a point at about 100 km up in the atmosphere that is the boundary between outer space and the atmosphere.

Auroras are caused when the solar wind (electrons, protons, and alpha particles released by the Sun) sufficiently disturb the magnetosphere (usually during a solar flare event). These particles collide with atmospheric atoms and molecules, resulting in the colorful display in the sky.

When auroras occur in the Northern Hemisphere, they are *aurora borealis*; in the Southern Hemisphere, they are *aurora australis*.

Figure 1.10. *This Aurora, seen in Alaska, was most likely the product of a geomagnetic storm that was caused by a coronal mass ejection from the Sun in the spring of 2015.*

Composition

The atmosphere consists of 78% nitrogen, 21% oxygen, about 1% water vapor, and small traces of other gases. Among these trace gases are greenhouse gases, which include methane, carbon dioxide, nitrous oxide, and ozone. Variations in the amounts of aerosols and ozone can affect the weather, climate, and quality of the air.

Several research programs exist to answer questions like:

- How is the composition of the atmosphere changing?

- How do these changes in the atmosphere affect solar radiation and the warming of the Earth?

- How do changes in composition affect air quality and various climates?

The main research programs that address these issues are:

- *The Upper Atmosphere Research Program* (UARP) studies the processes that control the concentration of ozone in the upper troposphere and stratosphere. Typical studies include studying the kinetics of reactions that directly and indirectly create and destroy ozone.

- *The Radiation Sciences Program* (RSP) works to improve satellite measurements to increase understanding of tropospheric aerosols and ozone, as well as their precursors and transformation processes in the atmosphere.

- *The Tropospheric Chemistry Program* (TCP) works to develop an understanding of how radiatively active gases, aerosols, and clouds scatter and absorb radiation from both the Sun and radiation emitted from the surface of the Earth.

- *The Atmospheric Composition Modeling and Analysis Program* (ACMAP) studies oxidation efficiency and air quality in the troposphere, including how polluting aerosols affect the properties of clouds and the increasing depletion of ozone. They focus particularly on long-term trends of atmospheric composition based on information received from weather satellites and ground-based measurements.

Structure

There is an inverse relationship between altitude and air pressure and density. That is, the higher in the atmosphere one goes, the lower the pressure and the lower the density. Temperature, however, may increase in some regions of the atmosphere or even stay constant. The top of all layers is generally warmer, as the warm air will rise and sit on top of the cooler air. The air there is very thin, meaning that there is much space between individual atoms of gas.

However, the closer one gets to the surface, the closer the molecules of gas get to one another. This dense cloud of gas can catch the high-energy *x*-rays and ultraviolet radiation coming from the Sun, which makes the molecules warmer.

The denser the layer, the greater its capacity to retain heat. Therefore, layers closest to Earth will be the warmest, and the layers furthest away will be the coldest. Despite the warmer temperatures of layers closer to Earth, they would still feel cold by human standards. This is because the molecules in the layer are too far apart to interact with properly and heat skin. The atmosphere consists of five layers, each bound by a "pause." Each of these pauses is where the greatest changes in temperature, density, movement, and chemical composition occur.

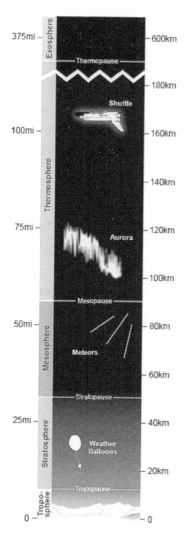

The exosphere is the coldest and outermost layer of the atmosphere. It is where atoms and molecules escape into outer space. This layer is where satellites orbit the Earth, around 600 kilometers above the surface.

The thermosphere is known as the upper atmosphere and is between about 85 and 600 kilometers above the Earth.

Along with space shuttles traveling in the thermosphere, auroras form here and can be seen from the Earth's surface (imagine those bright lights seeming so close but being more than 96.5 kilometers away). It is by far the thickest layer of the atmosphere.

The mesosphere extends from about 50 to 85 kilometers above the Earth's surface and has a temperature of approximately –15 degrees °C at the bottom of the layer.

The gases present in the mesosphere are close to one another and "thick" enough to slow down meteors traveling into the atmosphere; here, they burn up and leave fiery streaks that can be seen in the night sky (popularly called "shooting stars"). Both the mesosphere and the next layer down (the stratosphere) compose the middle atmosphere.

Figure 1.11. *Layers of the atmosphere and several phenomena that occur across it.*

The stratosphere is found between 50 kilometers above the Earth down to anywhere from 6.5 to 19.3 kilometers. This layer contains approximately 19% of all gases in the atmosphere

and is where weather balloons are deployed to take readings. The warm air sits at the top of the layer, which prevents convection, as there is no upward movement of the warm air.

The troposphere is also called the lower atmosphere. It is where all weather occurs. The layer begins at the surface of the Earth and extends anywhere from 6.5 to 19.3 kilometers high. The height of the troposphere varies with location; at the equator, it is about 19.3 kilometers high, but at the poles, it is a mere 6.5 kilometers.

Weather and climate

Weather describes the state of the atmosphere and the degree to which is it cold or warm, calm or stormy, dry or wet, and so on. Most weather phenomena that occur take place in the troposphere, the lowest layer of the atmosphere.

Weather is a measure of constant precipitation and temperature activities across the globe, whereas *climate* measures atmospheric conditions over some time.

On Earth, common weather phenomena include rain, wind, clouds, snow, fog, dust storms, and thunderstorms. Weather events that are less common include natural disasters such as typhoons, tornadoes, and hurricanes. All these events occur due to the differences in air pressure between two places.

Precipitation is defined as a water particle, either liquid or solid, that falls from the sky to the Earth. There are four main types of precipitation: rain, snow, sleet, and hail. Drizzle, snow pellets and snow grains are less common forms.

Rain is the most common type. Any drop larger than 0.0508 centimeters in diameter is considered a raindrop. Raindrops smaller than that together constitute a drizzle, a type of precipitation with uniform drops that are very close together.

Drizzle will appear to float while following the current of air. Unlike fog, drizzle will be heavy enough to fall to the ground. Both drizzle and fog often occur together.

Ice pellets, often called sleet, are translucent or transparent pellets of ice, which are normally round grains of ice. These are formed by raindrops that have frozen or snowflakes that have melted and then refrozen.

Hail is precipitation in the form of small pieces or balls of ice that fall separately or frozen together in large, irregular clumps. It is most often associated with thunderstorms, and it is considered a hailstone if it is 0.635 centimeters or larger. Larger-sized hailstones (reaching 2.54 centimeters or larger) are a sign of a severe thunderstorm.

Small hail, called snow pellets, are opaque, white grains that are round or cone shaped. They are not considered true hail because they are smaller than 0.635 centimeters.

Snow is composed of ice crystals that are branched and form six-pointed stars.

Snow grains are essentially frozen drizzle and are very small, opaque, white grains of ice.

Atmospheric circulation and the Coriolis Effect

Atmospheric circulation is created by masses moving in either a vertical motion (warm air rising and becoming buoyant) or horizontal motion (air flowing from high-pressure areas of dense compression to low-pressure areas of less density, creating wind). Sea breezes are an example of both forces interacting with one another. The *Coriolis force*, however, causes winds that move long distances to curve due to the Earth's rotation.

Every point of the planet rotates around the axis once every 24 hours at varying speeds; Air on the equator moves at 1,700 kph, while air at 60 degrees latitude moves at 850 kph since it is closer to the Earth's spinning axis.

Due to the Earth's rotation, *angular momentum* is created, which is the energy of motion that defines how an object moves around a reference point. This angular momentum is the product of an object's mass, velocity, and distance from the reference point, and it is conserved as an object moves on the Earth. Therefore, if its radius of spin decreases (as it moves from low latitude to high latitude), its velocity must increase.

French scientist Gustave-Gaspard Coriolis discovered the Coriolis force while he was studying why shots from long-range cannons repeatedly fell to the right of the target. Since the Coriolis force only affects long-range masses, it is not apparent in local weather patterns, nor does it make the water drain in different directions in the different hemispheres. It does, however, cause winds in low-pressure weather systems (such as hurricanes) to rotate and spiral.

The air starts to move in the typical high pressure to the low-pressure gradient, and, as it travels, the Coriolis force causes it to bend and create a state of geotropic flow where the pressure gradient force and the Coriolis force balance exactly. When this occurs, the air no longer moves from high-pressure zones to low-pressure zones but rather travels a course parallel to isobars (the notations used on weather maps to show changes in pressure). From this, a pattern forms where the air around low-pressure regions in the Northern Hemisphere is directed to the right and rotates counterclockwise around the system.

Conversely, the air in regions of high pressure in the Northern Hemisphere is directed away from the high-pressure force, producing a clockwise rotation.

Winds within one kilometer of the ground, however, are deflected toward low-pressure gradients because the friction with objects on the ground slows them down. As this air spirals into low-pressure areas, they rise the center of the convergence and begin to cool, creating condensation, clouds, and rain.

Some air parcels spiral away from the high-pressure areas, and rather than rise, they flow toward low-pressure areas, causing air to descend from above to maintain barometric balance – producing warm, sunny weather.

In 1735, English meteorologist George Hadley offered the first attempt to explain how weather patterns produce a circulation of the atmosphere. Hadley believed global scale circulation was simply a larger version of local systems, such as the one pictured above in Figure 1.12.

While his model was accurate in many respects (because the Earth is differentially heated, buoyancy develops at low latitudes, and mass is moved towards the poles, creating pressure gradients), Hadley's circulation pattern terminates at a latitude of 30 degrees.

The strength of atmospheric circulation is based on the dynamic of the balance between motions caused by differential heating and friction that slows the winds.

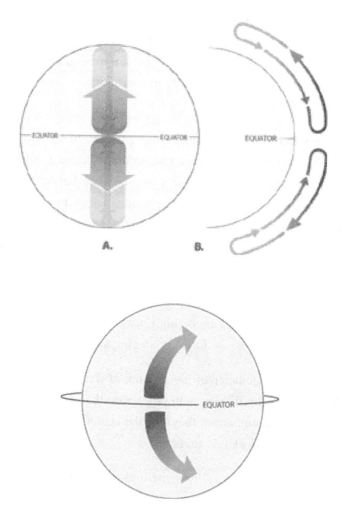

Figure 1.12. *If the Earth did not rotate on the angle of its axis, the air in the atmosphere would only circulate between the equator and the poles in a simple back and forth method. However, because the Earth does rotate on an axis, the circulating air is deflected left towards the Southern Hemisphere, and right towards the Northern Hemisphere. This deflection is termed The Coriolis Effect.*

Atmosphere-ocean interactions

The atmosphere and oceans exchange and store energy in the form of heat, momentum, and moisture. Oceans absorb heat more effectively than ice and land surfaces and store heat much more efficiently than land.

Coastal regions can remain more temperate areas, as oceanic heat is released at a slower rate than terrestrial heat. Changes in the balance of energy between the atmosphere and the oceans play a critical role in the Earth's climate change.

The currents of the oceans are affected by variations in the circulation of the atmosphere. The force of the wind drives currents along the surface of the ocean. The frictional drag of the wind moving across the surface of the ocean creates a current. This wind movement churns the water within a few meters below the surface. Below this mixed layer is the *thermocline*, a thin area of rapidly decreasing temperatures.

The ocean below the thermocline has its patterns of circulation, which are dependent on both salinity and the temperature of the waters. Vertical movements of water through and below the thermocline allow heat to be stored in the deep areas of the ocean and later released back into the atmosphere when recycled to the surface.

Oceans can affect the weather and atmospheric conditions due to their ability to store enormous amounts of moisture and heat. For instance, a tropical storm may form over a warm ocean, which will supply the necessary energy for typhoons and hurricanes to move and grow powerful and destructive. Winter storms that carry precipitation to large portions of the western United States begin their formation in the North Pacific Ocean.

ENSO

El Niño-Southern Oscillation (ENSO) is one of the longest studied and most important phenomena that occur on planet Earth, and it describes the natural year-to-year changes in the atmosphere and waters of the tropical Pacific.

The name *El Niño* (the boy child) was given because it typically occurs in December around the celebration of the birth of Jesus Christ. Even before 1900, scientists had begun studying ENSO events. Peruvian fisherman, back as far as the 1600s, had observed the warming of ocean waters off the coast of South America and the resultant impact it had on their fisheries.

El Niño can lead to large changes in sea-surface temperatures, sea-level pressure, winds, and precipitation (not just in the tropics but across the entire globe). An El Niño state takes place when the eastern and central equatorial Pacific sea-surface temperatures are much warmer than usual. El Niño typically results in much milder winters in North America. There is also an opposite pattern that exists called La Niña, which is a part of the ENSO cycle.

La Niña conditions occur when the eastern and central equatorial Pacific Ocean waters are cooler than normal (Figure 11, bottom). Though not always the case, a La Niña event typically follows an El Niño event. Both La Niña and El Niño last roughly eight to twelve months. Longer durations are uncommon but do occur occasionally.

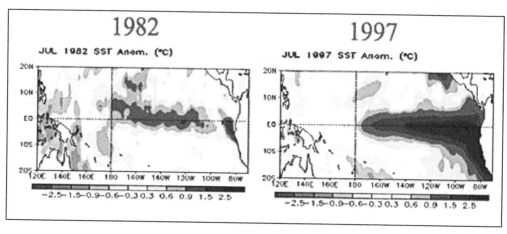

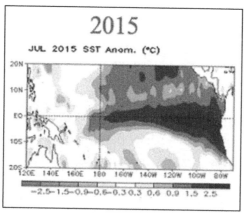

Figure 1.13. *Comparison of water temperatures in three El Niño events. In the 2015 event, notice the incredible increase in temperatures farther north in the Eastern Pacific. It is unclear how this anomaly will affect weather patterns around the globe.*

El Niño, A Temperature Anomaly

The Pacific Ocean receives more sunlight than any other area on Earth. Much of the sunlight it receives is stored in the ocean waters as heat. Normally, the Pacific trade winds blow from the east to the west, pulling the warm surface waters westward where they accumulate and pool to the east of Indonesia. At the same time, the colder waters from the deep Pacific Ocean rise to the surface, creating an east to west temperature gradient along the equator that is called a *thermocline tilt*.

Tradewinds often lose strength with the beginning of springtime in the Northern Hemisphere. As a result, less water is forced westward, and water in the eastern and central Pacific begins to heat up by several degrees Celsius, causing the thermocline tilt to diminish.

This balance is again maintained when the trade winds are replenished by the summer monsoon that occurs in Asia. In some instances, for reasons not fully understood, the trade winds will not replenish, or there is a reversed direction of wind current that travels from the west to the east. When this occurs, the ocean may respond in many ways. Warm surface waters that had pooled off the coast of Indonesia will begin to move eastward.

Beneath the surface of the ocean, the steep thermocline along the equator will flatten, as the warm surface waters will prevent the cool, deeper waters from upwelling. The result is the warming up of the large central and eastern Pacific regions into an El Niño. These waters, on average, will warm up by 3-5 °F but may warm up by more than 10 degrees in some areas.

As temperatures increase, the water in the east expands, causing sea levels to rise a few inches to as much as a foot. In the western Pacific, however, sea levels will drop, as much of the warmed surface water begins flowing to the east.

During an El Niño event in 1983, the drop in sea level was so severe that some corals were exposed and destroyed.

Notes

Global Water Resources and Use

The discussion of global resources and their use often starts with assessing the amount of water in various places on Earth, such as in glaciers, oceans, lakes, rivers, and groundwater. However, this discussion of inventories and reserves of water is a little off because, unlike fossil fuels, water is a renewable resource. Water from rivers and streams is continually replaced through a process called the *hydrologic cycle.*

To achieve sustainable use of water across the globe, emphasis must be put on the flows and fluxes of water. While knowing the amount of water in a lake or aquifer is useful, it will not be sustainable for humans if that water is not being replenished or if human consumption is faster than the rate of replenishment.

Water resources are not evenly distributed around the globe, and neither are the people who use them. The United Nations Environment Program has assessed human activities over the past two decades and found trends that reveal the following:

- Freshwater resources are not evenly distributed across the globe, and much of the water resources are located far from human populations. Several of the globe's largest river basins run through areas with very small populations.

- Agricultural uses of water account for nearly 70% of the total global consumption of water, mainly through the irrigation of crops. Industrial users account for about 20%, while 10% is consumer use.

- Groundwater accounts for nearly 90% of the world's available freshwater resources, and approximately 1.5 billion people depend on groundwater as their primary source of drinking water.

- Estimates show that two out of three people in the world can expect to live in a water-stressed area by the year 2025. Currently, about 450 million people in 29 countries are suffering from water shortages.

- Sanitation and clean water remain as two major issues in many locations around the world, with 20% of the world's population suffering from no access to clean water.

Water-borne diseases from unsanitary drinking water, often polluted with fecal matter, remain a major cause of illness and death in developing countries.

It is estimated that polluted water affects the health of 1.2 billion people globally and contributes to the 15 million child mortalities each year.

Freshwater & saltwater

Icecaps and glaciers make up an astounding 10% of the world's landmass and are primarily found in Greenland and Antarctica. They contain around 70% of the world's freshwater but are located far from human populations and are not an ideal or readily available resource for human use.

According to the *United States Geological Survey* (USGS), about 96% of the planet's frozen freshwater sits at the North and South poles, with the leftover 4% creating over 550,000 kilometers of glaciers and ice caps on mountains.

Groundwater is the most readily available and abundant source of freshwater on Earth, followed by reservoirs, lakes, wetlands, and rivers. Most freshwater lakes are in areas of high altitude, with about 50% of the planet's lakes located in Canada alone.

Many lakes, especially those in very dry regions, become salty through the process of evaporation. Reservoirs are human-made lakes, created by constructing physical barriers that reroute water to large basins to pool and be used for a variety of purposes. Wetlands include bogs, marshes, swamps, lagoons and floodplains and range in depth from 0 to 2 meters.

Current estimates place the total volume of water on Earth at approximately 1.4 billion km^3. About 30.8% of this water is stored underground in the form of groundwater. This makes up around 97% of all freshwater that is available for human use and consumption.

Only 2.5% of the volume of water on Earth is freshwater, and nearly three-quarters of this freshwater is frozen, covering most of the Arctic and Antarctic regions.

Freshwater rivers and lakes comprise about 0.3% of the world's freshwater supply, while the total usable freshwater supply for humanity is less than 1% of all freshwater and only 0.01% of all the water on planet Earth.

Where is Earth's Water?

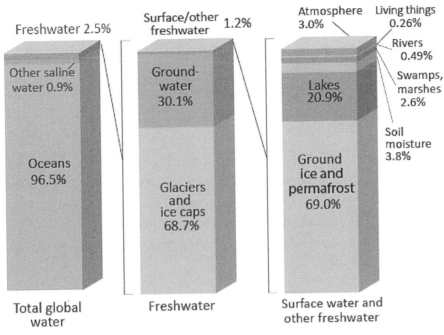

Figure 1.14. *The first column shows the division of fresh and saltwater sources on Earth. The second looks at the division of freshwater, while the third takes yet a closer look at surface water and other freshwater sources.*

Ocean circulation

Surface Ocean Currents

Ocean currents are due to the horizontal movement of seawater. Wind circulating above surface waters drive the currents, interacting with evaporated water and the *Coriolis Effect* (produced by the Earth's rotation). Stress between the ocean and the wind causes water to circulate in the same direction as the wind.

Currents can be temporary, affecting only pockets of regions. Others can be permanent and extend over large distances horizontally. Large ocean currents are contained within landmasses bordering the ocean basins. These continental borders force currents to form in a circular pattern, producing a gyre.

A *gyre* consists of four types of currents joining: two east-west currents (which form the top and bottom borders of the gyre) and two boundary currents that orient north and south. Each ocean basin contains a gyre, and the currents within these gyres are driven by atmospheric flow produced by subtropical pressure systems. The direction of flow within these gyre currents is due to wind circulation manipulated by the Coriolis Effect. Boundary currents are important in playing a role in redistributing global heat.

Surface Currents of the Subtropical Gyres

In all ocean basins, there are two west-flowing currents: The North and South Equatorial Currents. These currents can penetrate 100-200 meters below the surface of the ocean and flow between 3 and 6 kilometers daily. Flowing to the east is the Equatorial Countercurrent, which consists partly of a return of the water carried west by the North and South Equatorial Currents. During El Niño, this current strengthens in force in the Pacific Ocean.

Western boundary currents flow from the equator to high altitudes and are warm water currents with names associated with their locations: Gulf Stream (North Atlantic), Kuroshio (North Pacific), Brazil (South Atlantic), Eastern Australia (South Pacific) and Agulhas (Indian Ocean).

These currents are typically narrow and flow jet-like between 40-120 kilometers daily. The western boundary currents are the deepest of the ocean surface flows and penetrate 1,000 meters below the surface of the ocean.

Eastern boundary currents flow from high latitudes to the equator. These are cold water currents, which have names also associated with their location: Canary (North Atlantic), California (North Pacific), Benguela (South Atlantic), Peru (South Pacific), and Western Australia (Indian Ocean). These currents are typically broader than the western boundary currents and travel between 3 and 7 kilometers daily.

The eastward-flowing North Pacific Current and North Atlantic Drift of the Northern Hemisphere propel western boundary waters to the beginning point of the eastern boundary currents.

In the Southern Hemisphere, the South Pacific Current, the South Indian Current and the South Atlantic Current function in the same capacity.

Surface Currents of the Polar Gyres

Existing only within the Atlantic and Pacific basins in the Northern Hemisphere, the polar basins are pushed by counterclockwise winds. Westward-flowing currents forming the southern border of the polar gyres are the obverse of the eastward-flowing currents forming the northern border of subtropical gyres.

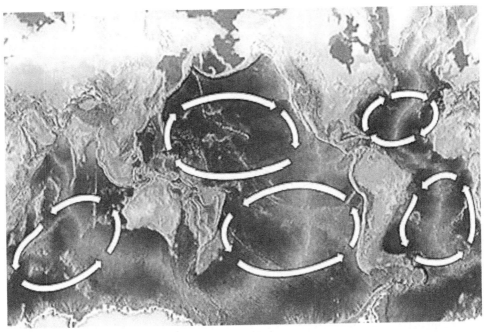

Figure 1.15. *This map describes the flow pattern of the five major ocean-wide gyres.*

Subsurface Currents

Currents also flow beneath the ocean's surface. These subsurface currents are much slower compared to surface currents and are driven by differences in seawater density, which differs due to salinity and temperature. Currents near the ocean's surface begin to travel deeper in the North Atlantic. This downward propulsion is due to high evaporation levels that cool and increase seawater salinity.

This cold, dense saline water sinks; this action takes place between Northern Europe and Greenland. This water then moves south along the coast of North and South America until it hits Antarctica. There, the cold and dense water travels east, joining another deep current that is created by evaporation and sinking between Antarctica and the southern edge of South America. During this eastward movement, the flow splits into two currents, one moving northward. In the North Pacific and the Northern Indian Ocean, these two split currents are drawn up from the floor of the ocean. The water warms up at the surface and then either forms another current that continues to flow at the surface (eventually getting back to the starting point of the North Atlantic) or creates a shallower flow that encircles Antarctica. It takes about one millennium (a thousand years) for the seawater to complete its circuit.

Agricultural, industrial and domestic use

Agriculture and food production consume the largest amounts of water, requiring nearly 100 times more than is utilized for personal needs. About 70% of the water drawn from groundwater and rivers is used in irrigation, while about 10% is used domestically and 20% in industry. Most of the water drawn for domestic use is returned to its sources, though wastewater is treated before being recycled. Since the 1960s, farmers and scientists have considerably improved global nutrition by providing more food per capita at lower costs. This was achieved using high-yielding seeds, plant nutrition, and improved irrigation techniques.

Crops are grown around the world but have very specific requirements for food and water no matter the location. Raising cattle for food also requires water — six to twenty times as much as is required to nourish cereal crops (grains). Almost every industry requires water to function during some part of the production process, such as those that produce wood, metal, paper products, gasoline, oils, and chemicals.

Figure 1.16. *Spray irrigation is a modern technique that waters crops much like a home sprinkler system (left); a low-pressure spray system that conserves water through the reduction of water loss via evaporation (right).*

Figure 1.17. *Drip/micro irrigation methods currently used that effectively water crops by using horizontal pipes that slowly drip water onto the soil. Using micro-irrigation systems can save water, increase crop yields, decrease the use of fertilizers, and decrease human labor demands.*

Water is also used globally to help produce electricity. Water may be re-routed from a river to power turbines that spin from the force of the onrushing water and produce electricity. This water is often recycled back into the river once it runs through the hydroelectric system. Dams are also used similarly to generate electricity.

Surface and groundwater issues

Surface water and groundwater have been and continue to be managed from various perspectives under several laws. Groundwater falls under the regulation of three laws, while the *Clean Water Act* or CWA (an amendment made in 1977 to *the Federal Water Pollution Control Act* of 1972) regulates surface water.

Both types of water are also regulated by state legislation designed to meet the standards set forth by the CWA and laws for groundwater.

Congress implemented the Clean Water Act to regulate the release of pollutants into surface waters across the United States. It gave the Environmental Protection Agency (EPA) the task of setting and maintaining standards for all pollutants in surface waters using water-based-technologies.

It is each state's responsibility to enforce the standards. Many states have developed independent agencies to do so.

CWA has three main goals:

- To eliminate all discharges of toxic contaminants in toxic amounts (no toxins is the goal);

- To eliminate all contaminants that are released into navigable waters by 1985; and

- To aid in the propagation and protection of wildlife, fish, shellfish, and recreation in and on water sources so that the water may be swimmable and fishable.

Before the early 1980s, most people thought that soil and rock filtered unwanted elements from groundwater and that the only concern regarding groundwater was discovering greater and more efficient ways to remove it from the Earth. In 1978, those beliefs about groundwater drastically changed when President Carter declared that a small area near Niagara Falls in New York was under a federal emergency, as there were growing health issues associated with groundwater and contaminated soil. The area became known as "The Love Canal" site and was the first area where a federal emergency was implemented in response to a human-made issue.

The U.S. population quickly realized that removing and treating groundwater before use was a serious issue that needed attention and doing so in the following decades became a large but necessary expense. The issue was worsened due to very few scientists before 1980 who specialized in these areas or had knowledge about technologies capable of fixing the problem.

Regulations aimed at protecting and remediating groundwater have focused mainly on hazardous pollutants. The chemicals commonly found in groundwater are widely used, from treated lumber to the manufacturing of electronic equipment, the production of plastics, fuels, the production of food and cleaning supplies, etc.

Sewage from both animals and humans in groundwater is becoming an increasing concern. States have adopted various standards to aid in the remediation and protection of groundwater, and most have legislated health guidelines that state the maximum permissible level of sewage-based contaminants in drinking water (that have been set by the EPA).

A few states have accepted "zero tolerance" standards, where the maximum permissible level of sewage contaminants in drinking water is the same as that found in uncontaminated water. This sets the bar high; scientists are finding that even the best technology available today is incapable of completely cleaning the groundwater from a site that has been contaminated.

Global problems

According to a report released in 2012, over 783 million people worldwide (about 11% of the population) do not have access to improved drinking water, such as water from public pipes, household connections, protected springs, or rainwater collections. Nearly 40% of all these people without clean drinking water live in Sub-Saharan Africa.

A lack of clean drinking water, or water for any use associated with personal health and hygiene, can cause an array of problems. Water-borne diseases due to contamination from fecal matter are particularly common, aggravated by the lack of good health care and medicine. In 2010, the United Nations General Assembly recognized that it was the right of every human on the planet to have access to clean and enough water for domestic and personal use.

The General Assembly stated that this water must be acceptable, safe and affordable (the cost must not exceed 3% of the household income) and that it must be physically accessible (the water source must be within 1000 meters of the residence and the total time for collection should not exceed 30 minutes).

The Millennium Development Goals Report in 2012 set certain goals to achieve in connection with the need for clean drinking water, and it has met them several years ahead of schedule.

The Millennium Development Goals include the following, among others:

- To eradicate extreme poverty and hunger;

- To reduce child mortality; and

- To ensure the sustainability of the environment.

Conservation

The conservation of water by avoiding waste and finding ways to use water efficiently are essential steps to ensure that there is enough water for the future. Freshwater is a finite resource, and it is up to every member of the planet to use water wisely.

In 2015, the state of California experienced some of the worst drought conditions in recorded history. Programs have been enacted that will give rebates to those who upgrade their toilet facilities and remove turf that must be watered to decrease water use. Though several laws have been enacted to aid in water conservation, there are simple day-to-day tasks and changes in lifestyle everyone can do to help conserve water.

To conserve water:

Bathroom: (where over half of all water used domestically is used)

- Turn off the water while brushing teeth or shaving;

- Take short showers and avoid using large amounts of water for baths. Restricting showers to five minutes will save over 1,000 gallons of water per month;

- When washing hands, turn off the water while lathering hands with soap; and

- Fix leaks from faucets and any pipes that may be loose.

Laundry:

- Wash loads only when they are full, or invest in a washer that allows water levels to be changed according to load size; and

- Consider investing in a high-efficiency washer, which can save over 50% in water usage and energy costs.

Kitchen:

- Wash only full loads of dishes in the dishwasher;

- Do not use running water to defrost foods; instead, thaw in the refrigerator; and

- Compost food waste instead of throwing it in the trash or using the garbage disposal.

Chapter 2

The Living World

Ecosystem Structure
- Biological populations and communities
- Ecological niches
- Interactions among species
- Keystone species
- Species diversity and edge effects
- Major terrestrial and aquatic biomes

Energy Flow
- Photosynthesis and cellular respiration
- Food webs and trophic levels
- Ecological pyramids

Ecosystem Diversity
- Biodiversity
- Natural selection
- Evolution
- Ecosystem services

Natural Ecosystem Change
- Climate shifts
- Species movement
- Ecological succession

Natural Biogeochemical Cycles
- Carbon
- Nitrogen
- Phosphorus
- Sulfur
- Water
- Conservation of matter

Ecosystem Structure

An *ecosystem* is a community of living and nonliving components functioning as a single system. This system works together to maintain life throughout the entire system.

Ecosystems can take many different forms and have differing levels of complexity depending on the varying living (*biotic*) and nonliving (*abiotic*) parts within the system. This integrated network of interactions is an *ecosystem structure*.

At its simplest, an ecosystem could be made up of one living organism in a non-living environment — for example, a plant inside a terrarium that provides adequate nutrients, water, and sunlight.

Ecosystems can also have a much more complicated structure. The most complicated ecosystem is, of course, the Earth — full of living organisms as well as nonliving components, such as water and sunlight, that work together to sustain life.

Figure 2.1. *Picture of the Earth from space.*

All ecosystems have the same basic structure. They include *primary producers*, which use sunlight to create energy and the organic compounds necessary for life. An example of a primary producer is a green fern, which grows out of the soil and uses sunlight to create organic compounds via photosynthesis.

Ecosystems also contain *consumers*, which feed on the primary producers to obtain energy and organic compounds.

An example of a consumer would be a deer that eats the grass and ferns, which contain the creative, organic compounds that were made via the process of photosynthesis. Consumers also break down organic components back into their inorganic parts, which can then be used by primary producers to repeat the cycle.

The interactions between producers and the organisms that eat them are called *trophic interactions*. Any time an organism eats a plant, a trophic interaction has occurred. *The food chain* is an example of a trophic interaction; energy is transferred from abiotic sources to primary producers to consumers and back again.

The food chain also helps transfer chemical compounds and sunlight to primary producers (e.g., grass) that use photosynthesis to make organic compounds, which are transferred to consumers of primary producers (e.g., a moose eating grass) and the consumers of the consumers of primary producers (e.g., a bear eating the moose).

Most ecosystems are complicated, and they contain many different abiotic and biotic components working together to sustain life.

Biological populations and communities

Biological populations are groups of individual organisms belonging to the same species that live within the same region concurrently. These populations can be described based on the entire population, as opposed to individuals within the population.

Populations can each have unique growth rates, mortality rates, age, and gender ratios. They also change over time, meaning that they may increase or decrease due to births, deaths or migrations.

For example, during a drought, many populations suffer significant decreases; conversely, when resources are plentiful, populations may increase quickly.

Figure 2.2. *Dry creek bed at Quivira National Wildlife Refuge due to severe drought in Kansas*

A population's ability to increase under perfect circumstances is called its *biotic potential*. That means that if a population were given access to perfect weather, constant and perfect food and water supply, ample shelter, and a lack of predators, the extent to which the population could feasibly grow would be its biotic potential.

Circumstances that may hinder population growth include such things as climate, food availability, and predation. When factors such as predators, limited food and water supply, inclement weather, or disease are added to the equation, a more realistic population size can be determined. The realistic population size that a particular habitat can support, given both the positive and negative attributes of its environment, is its *carrying capacity*.

There are two different kinds of biological populations, depending on their general response toward environmental capacity. *K-selected species* increase over time under good conditions but level off once they reach their carrying capacity. These species maintain their size by having fewer young, competing for resources, having longer life spans, and taking a long time to mature sexually. Examples of K-selected species are humans, whales, and elephants.

R-selected species increase rapidly and continue to expand to fill all available environments. Populations of these species maintain their size by having numerous young, shorter life spans, early sexual maturation, and little competition for resources.

Examples of r-selected species include most insects and bacteria.

The population sizes of both r-selected and K-selected species are dependent on many factors that can be divided into two different categories.

Density-dependent factors are factors that depend on how many organisms are in the area. These factors can include the availability of prey for the entire population or the number of mating pairs available.

Density-independent factors are factors that do not depend on the number of organisms in the area and include such things as changes in the weather or seasons. Both categories of factors serve to limit biological populations.

Populations interact with each other in different ways. For example, populations that eat only plants are *herbivores.*

Populations that eat other organisms are *predatory,* and their size depends on the amount of prey available.

Some populations work together in less violent ways, as in *parasitism*, when one organism benefits more than the other;

commensalism, when one species benefits while the other is neither helped nor harmed; and

mutualism, when both populations benefit.

An example of a parasite would be fleas on a dog.

An example of commensalism would be barnacles growing on a whale.

An example of mutualism would be beneficial bacteria within a human's intestines that aid digestion.

Biological communities are the total of all the different populations interacting within a common location.

For example, the moss, fern, and insect populations all living underneath the same tree constitute a single biological community.

Ecological niches

Just as people have their niches—as students, family members, or teammates—all species also have their niches. An *ecological niche* is a species' role within its environment. This role includes how it gets food and shelter, how it reproduces, and what it does to survive. It also includes how a species, through all its interactions with the biotic and abiotic factors in the ecosystem, help the entire ecosystem work. It is good for a species to have its unique niche, to reduce competition with other species in the ecosystem. When organisms share niches, they must compete for limited resources and opportunities, creating greater challenges to survival and success.

There are several ways to define a species' ecological niche, depending on what factor is considered. A *Grinnellian niche* considers a species' behavior about its habitat. For example, a mole's breeding, feeding, and protection all take place underground. Therefore, moles fill a niche as underground dwellers and impact the ecosystem from underground.

An *Eltonian niche* considers a species' behavior about its food preferences. For example, a niche of flying birds of prey that eat ground animals, such as mice, would have their Eltonian niche. Another Eltonian niche would be a niche just for animals that eat antelope, as they help regulate the antelope population.

A *Hutchinsonian niche* is more complicated. It considers a species' behavior in response to its environment (i.e., the resources and conditions that are required for the species to survive). For example, monarch butterflies lay their eggs on milkweed plants, and the caterpillars feed on milkweed plants once they have hatched.

A perfect niche, in which a species can use all the resources it needs to stay alive without competition, is called a *fundamental niche*. For example, monarch butterflies would have a fundamental niche if they had complete access to plenty of milkweeds, and no other organism ate the milkweed or destroyed it.

However, organisms typically must live alongside other organisms and, therefore, must compete with them for access to the different resources necessary to survive in their ecological niche. That means that monarchs must compete with other animals that eat milkweed, as well as humans who destroy milkweed during farming. This competition has led species to develop the ability to adapt.

Figure 2.3. *Example of a Hutchinsonian niche: the shape of this purple-throated Carib's bill fits the shape of the flower*

The environmental niche to which a species is most closely adapted is called its *realized niche*. A species' *adaptive zone* is the extent to which it can adapt to its environment. Monarch butterflies cannot currently adapt to lay eggs on different plants, so using other plants is currently outside of their adaptive zone. However, they may be able to adapt to life in captivity, which would then be inside their adaptive zone.

It is most important to understand that no two species can be in the same ecological niche at the same time, and both succeed. This would lead to too much competition for resources. As such, species adjust and adapt their ecological niches in response to other organisms within the ecosystem.

An example of this would be if monarchs had to move to a new location to find milkweed after a farmer had plowed over their previous milkweed patches.

Interactions among species

All biological species within an ecosystem must interact with each other. These interactions can take many different forms, and they are crucial to the success of the ecosystem as well as the survival of the specific species. These interactions can be between members of the same species and between members of different species.

One of the main interactions between species is *competition,* which occurs in the pursuit of resources, including food, water, mates, territory, etc. Competition most frequently occurs between members of different species; however, it can be between members of the same species as when multiple males in a mountain goat herd compete for access to females.

It is also possible for species to coexist in a very close relationship called *symbiosis.* This close relationship comes in many different forms of interaction, including *parasitism* (when one organism benefits while another is harmed), *mutualism* (when the species interact for mutual benefit), and *commensalism* (where one organism benefits and another is not significantly helped or harmed).

An example of mutualism is the good bacteria in the body that help maintain the body's pH, digestion, and other functions.

Figure 2.4. *Example of antagonism: the black walnut tree's roots release a chemical that harms nearby plants.*

Other interactions include *amensalism*, which occurs when a species harms another without receiving any benefit from it. For example, when a buffalo herd tramples their environment during migration, they are demonstrating amensalism. *Antagonism* occurs when one species benefits at the expense of the other. This usually means some predation occurs between species (e.g., when a deer consumes grass or a lion kills and eats an antelope).

Interactions between species and their impact on the biological population play a key role in the basic evolution theory via the concept of natural selection. This means that when species interact, one often comes out "on top" because of their superior adaptation to the situation and the environment. Also, certain individual organisms may succeed in their interactions with others more than other individual organisms due to their adaptation. Therefore, these interactions are important for "survival of the fittest," as well as for the sharing of genes during mating.

Keystone species

A *keystone species* is a species that has a disproportionally important impact on its ecosystem, considering the actual biological population of that species. These species play any number of critical roles in the success of the ecosystem as a whole and the survival of other species within the ecosystem.

If the keystone species suffers loss, other species will also suffer due to the loss of whatever important biological niche the keystone species is no longer able to fill. Therefore, understanding what species are "keystone" within an ecosystem is important when considering the impact certain events have on the ecosystem as a whole.

Keystone species may develop such status because of their position in a variety of different roles. For example, species can become keystone species based on their predation roles. A certain predator may eat enough of an invasive species to prevent it from overtaking the rest of the area's vegetation, dramatically changing the ecosystem landscape. This can be seen with birds that keep the mosquito population down.

Another example is when there are so many of a specific predator that they must eat a significant amount of another species, keeping that source of food diminished and preventing other species from using it as prey. The removal of any of these species would change the way the entire ecosystem is balanced, impacting every other organism in a watershed effect. Using the bird and mosquito example, if a virus decimated the birds that eat the mosquitoes, the mosquito population might increase exponentially, introducing mosquito-borne diseases to other organisms at a greater rate and harming their population levels.

Keystone species do not always need to be predators or even apex predators. They can be *mutualists*, performing tasks that benefit both themselves and other species. One such example would be flowers that feed other species with their nectar to have their pollen spread for procreation.

Figure 2.5. *The jaguar is an example of a keystone species.*

Keystone species may also be *engineers*. Their work to create and modify their shelters may create opportunities for other species. For example, bison prefer to eat plant life that grows in the land which has been previously excavated by prairie dogs. When beavers build dams in streams, they create entirely new ecosystems.

Altogether, there are many ways that even the smallest of species can become a keystone species, simply by the way they fill their biological niche and the way that other organisms depend on them filling that biological niche.

Species diversity and edge effects

Species diversity involves the number of different species that can be found within a specific ecological community. Two factors play a role in defining species diversity: *species richness* and *species evenness*.

Species richness refers to the numerical count of total species in the ecological community, while *species evenness* refers to the comparative proportion of species in the community and their level of equality.

For example, species richness would be the 500 squirrels in each ecological community, while species evenness would be those 500 squirrels in comparison to the number of chipmunks and other small rodents in the same ecological community.

Species diversity is heavily dependent on the specifics of the samples considered; different areas of an ecological community may have varying numbers of species depending on resources, different types of species, etc. Also, it is very important to define the boundaries of the ecological community when determining species diversity.

The places where these unique ecological communities come into contact are called *edges*, and the changes that occur to the populations in these specific edge areas are called *edge effects*. Many different ecological variables change at these edges, including temperature and moisture levels.

For example, certain kinds of plants grow at the edge between forests and suburban backyards (because of the specific moisture and sunlight levels present), and those plants are especially appealing to deer. The populations within these edges react and adapt to suit the area, such as when more deer come to an edge to feed.

Many edges and edge effects are humanmade. For example, the border between cleared farmland and walking trails, which are both humanmade, create edges to which the ecological communities on either side must adapt. When a hiking path is cut in a forest, more light and water can reach the ground in that area, changing the structure and niche of the ecological communities there.

The resulting increase in light and water may allow new plant life to thrive, attracting different animals to feed there than had originally frequented the area. This can create a trickle-down effect based on the change in the predator-prey structures within the community. This may ultimately affect the species diversity within a given ecological community, both by separating members of the community and changing the basic makeup of the species totals in the area.

Major terrestrial and aquatic biomes

Biomes are a distinct region with a community of plants and animals. There are many kinds of terrestrial and aquatic biomes. These biomes play extremely important roles in the adaptations of species and the definitions of ecological communities.

Terrestrial biomes are found on land. Mountains, forests, rainforests, and deserts are common terrestrial biomes, as they are all definable regions with specific kinds of plant and animal communities present.

The following are some less common terrestrial biomes.

Tundra is a high-latitude biome that is often covered in permafrost and populated with only low, shrub-like vegetation.

Taiga is a coniferous forest that occurs at high latitude.

Savanna lies between a grassland and a desert and is grassland with some scattered trees.

Chaparral is only found between 30- and 40-degrees latitude, mainly near the Mediterranean Sea, and includes tough shrubbery as a major plant feature.

A *scrub forest* is a response to drought. It contains small trees and thorns and has a long dry season.

An *ice cap* is a massive body of ice and snow that builds up over time.

Aquatic biomes are biomes that occur only in water. Lakes, ponds, rivers, streams, marshes, swamps, and coasts are all common examples of aquatic biomes.

Figure 2.6. *Farles Prairie in Ocala National Forest*

A *pelagic biome* is a far open ocean, excluding the ocean floor.

A *benthic biome* is a deep-sea habitat, often with little light and heavy pressure from the miles of water above.

A *reef* is a structure created by coral polyp skeletons that houses hundreds of species of fish.

An *oceanic vent* biome is an area where hot sulfurous water is pumped out from cracks in the seafloor where continental plates are pushed apart. These biomes house some very highly specialized species able to tolerate such adverse conditions.

Temporary pools are bodies of water that only exist for part of the year before draining and drying for the remainder of the year, forcing species to tolerate the lack of water for extended periods.

An *abyssal biome* is the true deep sea, often with the absolute absence of light and intense pressure from the water above.

Brackish water is saltwater.

Bogs are wetlands wrapped around a body of water where plant life flourishes.

Energy Flow

Ecological energy flow is the transfer of energy through the various levels of the food chain. All sources of energy on Earth originate from light energy from the sun. These energy sources are then transferred to primary producers that synthesize glucose from chemicals in the soil combined with the sun's light energy.

For example, grass makes glucose out of chemicals and light energy during the photosynthesis process. This energy is then hypothetically absorbed by the next level of the food chain (primary consumers) who eat the plants (primary producers). An example of this would be a rabbit eating the grass, thus absorbing the glucose that the grass contains.

Next, secondary consumers eat the primary consumers, hypothetically transferring that original energy again. An example would be a fox eating the rabbit. There may be other consumers after the secondary consumers, but generally, there are no more than tertiary consumers for reasons that will be discussed later.

The secondary consumers eventually die, and as their bodies decompose, nutrients are transferred back into the soil and water. After the fox dies, its body is decomposed by flies and maggots, eventually returning to the Earth as chemicals. These nutrients can later be used by primary producers in conjunction with the sun's light energy to create transferrable energy yet again, thereby completing a cycle of energy flow.

Every time energy is transferred from one level of the food chain to another via consumption (e.g., each time the original energy goes from the grass to the rabbit to the fox), approximately 90% of it is rendered unusable because it is either expended through heat or is found in indigestible food.

That means that 90% of the energy stored in the glucose molecules is unusable, and it is lost to the heat created during the transfer, or it is found inside the food that the animal can't consume, such as bones. Therefore, each consumer only gets 10% of the energy the previous consumer obtained. That is why there are rarely more than tertiary consumers in an energy flow chain; beyond secondary consumers, the amount of energy obtained by eating the food is not enough to meet a consumer's energy requirements.

The sections below describe the various processes that are included in this energy flow. Each process works together to transfer energy across all species within the same ecological community. In this way, energy is never truly lost but is transferred and eventually rendered unusable along the chain.

Photosynthesis and cellular respiration

Photosynthesis and cellular respiration are critical to the entire energy flow process because they are the first steps in the equation. These two steps are performed within primary producers (plants containing *chlorophyll*) to transfer light energy and nutrients found in soil into energy that is usable by plants and, eventually, usable by the rest of the food chain.

Photosynthesis is the first step in this two-step process. The goal of photosynthesis is to create a sugar molecule out of carbon dioxide, water, and energy. Carbon dioxide from the air and water from the ground are combined with *chloroplast* cells, or cells containing green chlorophyll pigment.

Light energy is then applied to the chemical reaction, which converts the carbon dioxide and water into a six-carbon sugar and oxygen gas to excrete back into the atmosphere. This six-carbon sugar molecule is *glucose*, and it is necessary for biological life as an energy source.

The energy is contained within the strength of the chemical bonds between the six-carbon molecules in glucose. It is stored in those bonds until the organism intends to use it as an energy source, at which point the carbon bonds must be broken to release the stored energy. This process is *cellular respiration* and requires oxygen.

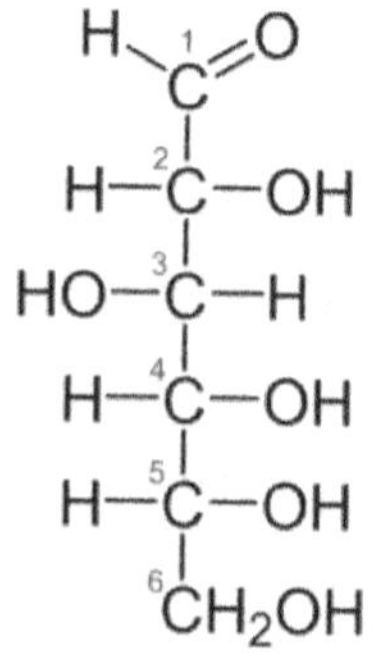

Figure 2.7. *Glucose*

Cellular respiration is a chemical reaction that occurs after successful photosynthesis, and it does not only in chlorophyll-containing organisms. All organisms use cellular respiration to change glucose into energy they can actively use. Once the glucose has been created by photosynthesis, it can be transferred via the consumption of the organism containing the glucose by a consumer organism. The strong molecular bonds found in the glucose molecular structure can then be broken to free energy within the mitochondria of all living cells.

Cellular respiration is an overall name for several processes to convert glucose to energy. The process used depends on whether oxygen is present during the reaction.

Aerobic respiration processes require oxygen, and *anaerobic respiration* processes do not. Aerobic respiration occurs in the mitochondria of cells by combining glucose and oxygen to obtain roughly thirty-six ATP (adenosine triphosphate).

The first step of this process is *glycolysis*, followed by the *citric acid cycle* and, finally, the *Krebs cycle*. Together the three steps take glucose and oxygen into approximately thirty-two to thirty-eight ATP (depending on the organism) that can be used by the cells for energy.

Anaerobic energy takes glucose without any oxygen to create a net of two ADP and lactic acid in animals or carbon dioxide and ethanol in plants. It does so via proton pumps in cell membranes.

Food webs and trophic levels

A *food web* is the pictorial representation of the natural ways organisms within an ecological population eat each other. It represents the ways that energy is transferred from soil, all the way to the last predator, and back to the soil during decomposition.

The following map shows the interrelated connections created by feeding patterns and allows scientists to see energy transfer and energy loss across *trophic levels*.

An example of a food web would be a pictorial representation of the previously discussed example of light energy being transferred to grass to a rabbit to a fox and then back to grass through decomposition.

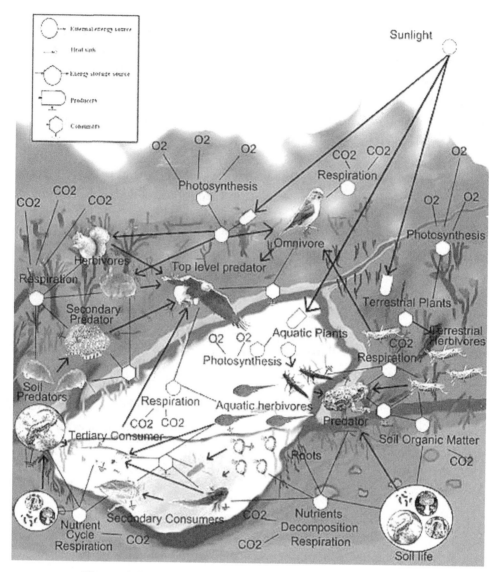

Figure 2.8. *Freshwater aquatic and terrestrial food web.*

Trophic levels are the positions that animals occupy in the food chain and, therefore, on the food web. They can be used to categorize animals based on their positioning.

Trophic levels are broadly divided into three groups:

autotrophs, or organisms like plants and algae that do not eat other organisms but transfer nutrients from the environment into the food they make themselves;

heterotrophs, or organisms that do not make their food but instead obtain energy by eating autotrophs and other heterotrophs; and

decomposers, which break down dead plants and animals and return their nutrients to the soil.

Examples of heterotrophs are cats, bears, and snakes, as these animals cannot produce their food and instead eat others for energy.

An example of a decomposer would be a fly, particularly its maggot larvae, as they feed on animal carcasses and return nutrients to the soil through their excrement.

Animals can be categorized into numbered trophic levels depending on where they fall along the food chain. These levels are:

Level 1, primary producers: plants and algae making their food;

Level 2, primary consumers: herbivores that eat the plants and algae;

Level 3, secondary consumers: predators that eat the herbivores;

Level 4, tertiary consumers: predators that eat other predators; and

Level 5, apex predators: predators that have no predators themselves (essentially the top of the food chain).

In a hypothetical situation, this could be exemplified by a level 1 shrub, followed by a level 2 antelope that eats the shrub, a level 3 hyena that eats the antelope, a level 4 crocodile that eats the hyena and a level 5 leopard that eats the crocodile. The leopard would be considered the apex predator in this food web and would occupy a level 5 trophic level.

Decomposers feed on the dead bodies of all trophic-level organisms, returning them to the nutrients that can be used by primary producers to start the process over again.

Energy is transferred along with the various levels of the food chain and can be described as *biomass*. Biomass is a descriptive term that covers the amount of energy available by consuming an entire trophic level, and it is a ratio of the energy available when considering the size of the organism consumed.

Biomass levels may seem skewed when considering certain animals with many inedible parts, such as tusks or hooves. However, it is a term used to measure the flow of energy across trophic levels indirectly.

Ecological pyramids

An *ecological pyramid* is a pictorial representation of the various levels in an ecological community, as previously described. There can be several different kinds of ecological pyramids depending on what facet of the ecological community is being considered.

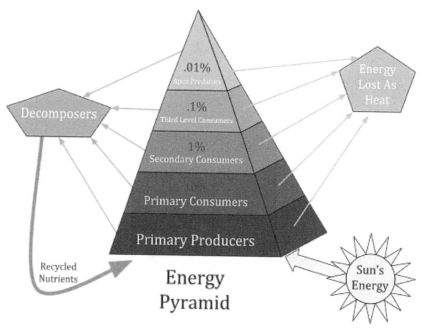

Figure 2.9. *Ecological energy flow*

An energy pyramid stacks the five trophic levels on top of each other and then attempts to quantify the energy transfer percentages. It begins with the producers on the bottom and follows the transfer of energy upward through the different trophic levels. It allows a scientist to look at the approximate available energy in an ecological community at any given time.

There can never be an inverted energy pyramid, as there must always be more energy at the lower levels than at the higher levels for there to be enough energy to make the ecosystem viable.

Whether the ecosystem is a forest ecosystem or a desert ecosystem, there will always be more initial energy available at the primary producer level than at the apex predator level.

Similarly, an energy pyramid of biomass tracks the biomass throughout the ecological community. Each layer presents a different pyramidal shape based on the different percentages of available biomass.

For example, in a stream ecosystem, the level of the main producers for the ecosystem — water plants — will be smaller than the primary consumer level, as there is vastly more "mass" associated with the insects and fish in the stream.

Biomass energy pyramids can be deceiving because, at higher energy levels, certain species may have more mass. These species, such as birds, have a greater mass ratio to the energy that is available to them at that level, making their biomass ratios skewed.

A numerical ecological pyramid shows the ratios of species at each trophic level. This type of pyramid can be many different shapes depending on the ratios of species.

One example of this is if an ecological community has more primary consumers than it does primary producers.

For example, in a community with only five or six species of plankton and algae but upwards of ten species of fish and birds, there would be a larger level for the primary consumers (fishes and birds) versus the primary producers (plankton and algae).

Figure 2.10. *North Sea phytoplankton bloom.*

Ecological pyramids function as a way for scientists to monitor the transfer of energy within an ecosystem in multiple ways. This allows them to understand the delicate balance between all the species and niches within an ecosystem, as well as what would happen if any single species was diminished in reference to the others.

Ecological pyramids help scientists gain a greater understanding of an ecosystem.

Ecosystem Diversity

Ecosystem diversity, or *ecological diversity,* is a description of all the various ecosystems within a region, but it can also apply to the various ecosystems found on the entire planet. Therefore, there are innumerable diverse ecosystems included in the concept of "ecosystem diversity."

Ecosystem diversity considers the different trophic levels, the complexity of the systems, all the different niches that the organisms fit into, and the differences in environments within the ecosystem.

As the largest scale on which to observe the diversity of life on the planet, this overall title must be divided up into subcategories. The different types of diversity are just as numerous as the ways it can be observed across the planet.

Some of these options include *biodiversity*, *species diversity,* and *ecosystem diversity*, but there can be many others. Biodiversity and species diversity are discussed in greater detail below. When looking strictly at the ecosystem, the many different biomes, ecological communities, and other aspects are all considered.

Evolution and natural selection attempt to explain the existence and development of such incredible diversity, specifically in organisms. This diversity is termed biodiversity.

Biodiversity

Biodiversity is the incredible variety of life on this planet. It can be divided into many different categories due to the numerous varied species to be considered. *Species* are a group of living organisms that can breed with each other.

For example, elephants are a single species because they can breed with each other, while elephants and crocodiles — though they may live within the same ecosystem — cannot be considered the same species because they cannot breed successfully with one another.

Biodiversity considers all the species present within an area; it is the number of different species in an ecosystem.

Biodiversity is greatest in warmer, wetter climates, as those areas are more conducive to great numerical variety in species. In such areas, the variety of species has been increasing over time.

Since biodiversity is such a vast concept, there are several ways to narrow it into subcategories. One of these is *genetic biodiversity*, or the many variations of genes that can exist within the same species. This includes all the different hair colors, eye colors, and face shapes, as well as negative genetic mutations.

Therefore, different breeds within a single species can show genetic diversity — e.g., all canines are the same species, as they can breed with each other, but they show a range of genetic diversity from Chihuahuas to Great Danes. Another subcategory of biodiversity is *ecosystem variation*, or the many different varieties of plant life that comprise biomes where organisms live. There can be marine (water) biodiversity, as well as terrestrial (land) biodiversity.

Biodiversity is responsible for the great genetic variation that exists on our planet, which supports survival and change. Without a great deal of genetic diversity, inheritable diseases become more prevalent, as demonstrated by the high rates of genetic disorders in closed communities like the Amish.

Genetic variation also allows species to adapt by letting the stronger, better-adapted organisms survive, reproduce, and pass their successful genes on to the next generation.

However, many things are occurring on our planet that threaten to limit biodiversity. Habitat destruction, poaching, pollution, and climate change all have the potential to limit biodiversity on the planet drastically.

Each time a species becomes extinct, a great deal of diversity is lost that cannot be recovered. That is why it is so important to conserve the wide biodiversity on the planet so it will continue into future generations.

Natural selection

Natural selection is the major driving factor in Darwin's theory of evolution. All species face different challenges and levels of adversity. The *theory of natural selection* says that certain genetic variations within individual organisms may better facilitate survival in the face of adversity. Individuals with certain traits or genes may have a better chance at surviving and reproducing, thus passing along their "superior genes."

For example, animals with longer necks may have had an easier time reaching food, and genes for long necks in giraffes were passed on to the next generation.

Over time, these genes become more prevalent in species through an increase in offspring and the possibility that those individuals with less successful genes are more likely to die before having the opportunity to pass on their genes. Certain species may begin to develop specializations for their specific niches, such as developing plumage that allows them to camouflage in their environment.

Over time, these specializations may become so developed that organisms possessing them to become new species. Animals with better plumage, for instance, may have found survival much easier than those without it and may become so genetically separate from their original species that they form a distinct species over time. That is the natural selection theory of biodiversity.

Figure 2.11. *Charles Darwin, English naturalist and geologist*

Specific changes through natural selection can be defined. *Positive selection* is natural selection that increases the presence of a certain genetic allele in the gene pool.

For example, the increase in the prevalence of a gene that provides immunity to mumps would be a positive selection.

Negative selection reduces the presence of deficient genetic alleles in the gene pool. An example would be the death of those individuals without immunity to mumps during a mumps outbreak, thereby reducing the existence of their "deficient" genetic alleles in the gene pool.

Changes can also be classified based on what part of an organism's life cycle the variation impacts, such as survival versus fertility. Changes that increase survival would include the neck growth in giraffes, as discussed above, as well as the immunity to mumps. Changes in fertility, however, would include natural selection for individuals who have increased fertile windows, allowing for greater reproduction opportunities.

Over time, the species may increase their fertile windows through the process of allowing individuals with longer fertility windows greater opportunity to procreate compared to those without them.

One of the most well-known examples of natural selection is in the antibiotic resistance of bacteria. Some bacteria may survive antibiotic assault because of mutations in their genetic structure, and, if they do, they can reproduce to create a new colony of bacteria that are no longer susceptible to the antibiotic while the susceptible bacteria die.

Natural selection contrasts with artificial selection, such as breeding for specific traits, which is seen in dog and horse breeding. Natural selection occurs due to the impact of genetic diversity on overall survival and reproduction, while changes due to artificial selection are controlled externally.

Evolution

Evolution is simply the change in heritable traits in populations over time. When combined with natural selection, evolution serves to change species' observable traits over a great deal of time, sometimes until they become entirely new species. This does not try to explain the creation of life but instead describes the way that species adapt until they change visibly in response to their environments.

These visible changes eventually accumulate to create branching species and increased biodiversity. This is an explanation for the vast biodiversity on our planet and a prediction that biodiversity will continue to increase over time.

As part of his *Origin of Species,* Darwin was the first to propose such a theory to explain Earth's biodiversity. He proposed that natural selection ensures that genetically superior organisms survive to reproduce and pass along their superior genes, even though changes in genes to create such superiority must come from DNA replication mistakes.

This means that some "mistakes" actually are of benefit to organisms, and over time those "mistakes" become more prevalent via natural selection and begin to transition species in new directions. The same DNA changes tend to become more prevalent in smaller populations due to the reduced number of breeding pairs.

To confirm the theory of evolution, scientists search for the fossilized remains of species that show a transition from one form to today's visible form of a similar animal. Many different fossilized organisms are "intermediaries" along the evolution chain.

These intermediaries show what animals looked like before they had successfully made it to the next state after the change. There are many different examples of transitional forms before man's development into *Homo sapiens*, exemplified by changes in the skeletal structure, transitioning to upright walking, etc.

Figure 2.12. *Shells and fossils*

Other things that scientists use for proof of past evolution include *vestigial structures* or structures within an organism that serve no known purpose and may be carryovers from previous organisms. The appendix in a human is a vestigial structure, as it serves no currently known purpose, and a human can survive without it.

Scientists also use *embryology*, or the study of the different stages embryos undertake during development, to find similarities to other organisms in an evolutionary context. Many organisms that are different at birth share similar characteristics when still in the embryonic stage, such as webbed hands, tails, or even gills.

Also, scientists consider structures, bones, and other things that are shared across different species. For example, arms and fingers share a remarkable similarity to wings of certain species, and scientists attempt to map evolutionary pathways between animals with wings and those with arms that share similar bone structures.

Regardless of a person's stance on evolution, species do change over time in response to their environments, and genetically superior individuals will have a better chance of reproducing. This can drastically alter biodiversity now and in the years to come.

Ecosystem services

Ecosystem services are the various ways that people benefit from their given environments. Organisms derive many different benefits from their environments, as environments are quite literally the source of all support that organisms must survive.

These environmental services interrelate to create niches, support biodiversity, and support evolution and natural selection. Studying the different environmental services offered allows scientists to observe ecosystems and the way they function to support overall life.

Ecosystem services are generally placed into four different categories:

provisioning, services for the creation of products for use;

regulating, services that regulate ecosystem processes;

supporting, services that are required for all other services to work; and

cultural, services such as religion and spirituality.

An example of provisioning services is the ecosystem's provision of food and water in support of plant life and streams.

An example of regulating services is pest control, such as seasonal change killing off mosquitoes every winter.

An example of supporting services is the water cycle and water purification.

Examples of cultural services include a beautiful part of a reservation used for Native American religious ceremonies and a whitewater rapid providing recreational opportunities for people to kayak and raft.

Other examples of ecosystem services include minerals and crop pollination via bees and other insects. Any benefit that humans receive directly from the environment that they did not engineer themselves is an ecosystem service, and even a brief overview of all these services shows how much humans receive from the planet's natural resources.

These ecosystem services offer great economic benefit to humans by taking care of problems that would be incredibly expensive or impossible to replicate by hand. This makes it extremely important for humans to support and preserve ecosystem services for use by current and future generations.

There are presumed to be some ecosystem redundancies built into these services, meaning that several species or systems may provide the same ecosystem service, but even so, it is important to support and protect these ecosystem services.

By protecting the species and systems that create ecosystem services, there is protection for the continued ecosystem balance of the Earth.

Natural Ecosystem Change

Ecosystems change over time in response to various factors, just as organisms change in response to their environments and external forces.

However, the pressures that force ecosystems to change are not as simple and as widely understood as the processes that force organisms to change.

There is significant research on how humans have both positively and negatively impacted all levels of ecosystem change by affecting ecosystem services. There are also natural processes that have a significant impact on the ways that ecosystems change over time.

Climate shifts

Climate shifts are changes in weather patterns across the planet that last for an extended period. Therefore, short-term weather shifts, such as an exceptionally cold winter or rainy season, are not indicative of climate change; instead, the weather pattern shift must be sustained over a long period — decades or more — to qualify as actual climate change.

Climate change can be caused by a variety of factors, not all of which are humanmade. Potential causes include changes in levels of solar radiation, shifting plate tectonics, volcanic eruptions, space debris impact, and various actions caused by living organisms on Earth.

Each of these things has the potential to cause catastrophic cooling, as in the case of ash remaining in the atmosphere for long periods, or catastrophic heating, as in increased solar radiation due to the ongoing loss of the protective ozone layer. Such catastrophic heating or cooling could have a cascade effect on the planet's plant and animal life by impacting environments, including significantly changing water levels worldwide or increasing the severity of weather systems, such as hurricanes and typhoons.

Figure 2.13. *Satellite photograph of clouds caused by the exhaust from ship smokestacks.*

The Earth has historically undergone many different stages of climate change, each caused by different factors. In all cases, climate change has balanced out enough to allow life to continue to flourish on the planet. The various ice ages are one major example of past climate change.

Each climate shift has the potential to drastically change ecosystems by increasing or decreasing water tables, changing vegetation, altering the structural landscape, or causing mass extinctions that change how the ecosystem is maintained.

Scientists have studied past climate shifts to understanding how they occurred and how they may be managed in the future. Scientists study these changes by taking deep ice-core samples and reviewing sediment layers and past sea levels to observe changes over time.

Climate change is a controversial topic, mainly because of differing opinions on its causes. Climate change can occur as a result of both natural and human-made actions. Also, past global climate shifts have occurred in the absence of man-directed activity and have stabilized to allow life to continue on the planet. Regardless, it is important to understand how human actions could potentially impact the stability of the planet's climate. Actions that have occurred since the Industrial

Revolution, including the decrease in the ozone layer and increases in certain chemical concentrations in the air, water, and soil, have had major impacts on the overall climate shift. It is important to understand the human impact on the climate and mitigate what humans do as much as possible to protect Earth's complicated ecosystem balances.

Species movement

Whenever climates or environments change, species must adapt to survive in their new environments. Sometimes these shifts are not especially challenging, and the organism can easily adapt to the new environment. At other times, species find that their current environment is no longer suitable for their survival for various reasons, and they must move to find an environment in which they can survive. This is *species movement*.

This movement can be as simple as a routine and necessary *migration* from the environment to the environment. Animal migrations are predictable events in response to such environmental changes as weather or temperature change, the need for better breeding locations, or changes in food availability.

An example of such an animal migration is Canadian geese migrating south for the winter or bald eagles migrating north to seek preferred temperature ranges. These species' migrations are typical responses to regular environmental change and therefore are not indicative of unique changes to the environment.

In contrast, an unplanned move can be disastrous for the species. Some species may not have any preparation for a new environment if their old environment is destroyed by an unexpected disruption (e.g., volcanic eruption, wildfire, logging).

Some species may be so highly specialized that no other location would fully support their survival, such as marine animals that must live underwater and cannot move elsewhere if their current water becomes polluted or too warm for them to survive. Others may find themselves losing their competitive edge in the new climate.

If another species can better adapt to the adversity and begin to take over resources, they will be more successful in surviving.

As a result, these species must either adapt to the changes in their current environment or risk extinction. In some cases, this is the natural process of change in ecosystems. However, it can also be indicative of negative climate shifts killing off important species, including keystone species, due to the disruptive changes made to the animals' environments.

For example, polar bears are currently dying out due to a decrease in ice cover, the increase in water tables resulting from melting glaciers, and a reduction in the presence of prey. These changes mean that there are fewer predators in these ecosystems, which fundamentally changes the entire energy pyramid for the environment and has other long-reaching consequences on the whole ecological community.

The changes contributing to the loss of polar bears, a major player in a major ecosystem, have the potential to impact other ecological communities in a trickle-down effect negatively. As such, humans need to be aware of the causes behind species movement and how it is tied to climate change and other unintended — often negative — consequences.

Ecological succession

Ecological succession is the observable change in the species present in an ecological community over time as it responds to changes in its environment.

Ecological communities start with relatively few plants and animals that increase and adapt over time until the ecological community grows strong and stable with varied populations.

These populations begin to respond to each other as they interact and grow, creating an interrelated ecosystem. Succession can also be used to describe the process that environments undertake to "bounce back" from disasters, such as volcanic eruptions, earthquakes, and forest fires.

Figure 2.14. *Succession after a wildfire in a boreal pine forest.*

There are several types of ecological succession. The first kind of ecological succession is *primary succession*, which is the creation of a new community in a previously uninhabited area. This happens relatively infrequently, as it is rare to find uninhabited locations left for species to newly inhabit.

Secondary succession is the growth of a new community in a location after some traumatic event, like a forest fire.

Cyclical succession follows predictable patterns based on routine climate and environmental change, like the ecological bounce back that occurs every spring after a severe winter.

Autogenic succession is based on the changes in the soil that are brought about by the addition of new plant species that allow additional species to join the new community.

An example of autogenic succession would be if a new shrub that fertilized the surrounding area was introduced and nearby plants flourished in response to the positive change that that shrub brought to the community.

Allogenic succession is caused by external impacts and not by what the surrounding organisms create.

An example of allogenic succession would be if an area's soil improved due to water runoff from a nearby stream, changing the makeup of the soil, and improving conditions for the area's vegetation to thrive.

Succession ends when the ecosystem has reached an equilibrium between the environment and the organisms populating it. This is called the *climax*. Ecological communities grow first with a strong emphasis on r-selected species that reproduce quickly but are eventually replaced by K-selected species that live longer and reproduce infrequently. This helps to stabilize the community until it reaches the balancing climax.

However, a climax is nearly impossible to attain, as there is almost always some disruption in every ecological community. These disruptions are followed by a "bounce back" in one form or another, so the ecological community is never fully balanced.

Additions and removals to the ecological community force the community to be in a near-constant state of flux, even if the disruption is not severe, so a perfect climax is nearly impossible to attain.

Patterns and time frames for ecological succession are important for scientists to observe in response to ecosystem loss and today's shifting climate. This information allows scientists to understand what would be required for a community to rebuild following major damage.

Natural Biogeochemical Cycles

A *biogeochemical cycle* is a process by which chemicals cycle across the Earth and are made accessible at various levels. These levels include the water system (hydrosphere), the soil (lithosphere), and the different layers of the atmosphere. Vital chemicals are passed across the Earth and from organism to organism via biogeochemical cycles, which are necessary to sustain life on the planet.

A predominant example of a biogeochemical cycle is the water cycle, which transitions H_2O through various states in the atmosphere in a cyclical fashion. However, other chemicals are involved in biogeochemical cycles as well. These biogeochemical cycles recycle and continuously move chemicals throughout the environment to make them available for all organisms.

There are several different kinds of biogeochemical cycles, depending on the chemical being transported and the locations to which it cycles.

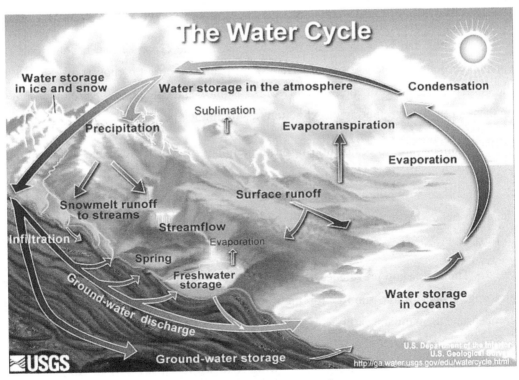

Figure 2.15. *Water cycle*

Carbon

The carbon cycle is necessary to sustain life on Earth, and it is the process by which carbon is exchanged across Earth's entire atmosphere. Carbon (C) is an organic compound found in every living creature on Earth. Within the "carbon cycle," many levels are depending on location, such as the cycling of carbon above the ocean and over land or the geological carbon cycle within rocks.

All these cycles combine into the overall carbon cycle, which — along with the water cycle and the oxygen cycle — is one of the processes that make life on Earth possible.

The carbon cycle is generally described based on where it occurs. Carbon dioxide leaves the atmosphere through plants via photosynthesis, then enters the terrestrial and oceanic biospheres. It can also enter the water cycle through dissolving into precipitation in the form of carbonic acid, a key component of "acid rain."

In the terrestrial biosphere, carbon is passed through various organic compounds when organisms consume each other, as well as when their bodies decompose. It can easily return to the atmosphere when organisms exhale carbon dioxide back into the air.

Carbon can be used during photosynthesis in the ocean or can become part of calcium carbonate in shells or sediment. A significant amount of carbon is also stored in the Earth's lithosphere, or rocks (mainly in limestone), and can be held there indefinitely or shed via runoff, weathering or extraction (as in mining operations).

Humans have a major impact on the amount of carbon available in the carbon cycle. For example, deforestation reduces the number of trees able to pull carbon from the atmosphere during photosynthesis. Air pollution also introduces additional carbon to the atmosphere that would not naturally be present, which can overwhelm plants' ability to filter it out.

The increasing prevalence of fossil fuel, exhaust, and fertilizer add carbon to the atmosphere and water runoff, forcing the rest of the carbon cycle to attempt to compensate to maintain carbon balance. These problems are currently having a large negative impact on the Earth's climate, as this increase in carbon is changing the structure of Earth's atmosphere and allowing it to warm on the surface, impacting all living things.

Nitrogen

Nitrogen (N) is essential for life on Earth. Most of the atmosphere is made up of nitrogen. It is part of all amino acids, which are the building blocks for proteins and, therefore, part of the structure of all living cells. It is necessary for all living organisms to have access to nitrogen to produce and use vital amino acids and proteins.

Several processes in the nitrogen cycle make the different forms of nitrogen available for use by living organisms on the planet. These processes include nitrogen fixation, assimilation, ammonification, nitrification, and denitrification.

Nitrogen fixation turns atmospheric nitrogen into a form that can be used by plants via symbiotic bacteria that transfer atmospheric nitrogen into ammonia and carbohydrates.

Assimilation is the process by which plants can absorb nitrogen from the soil, which is then transferred to the animals that consume the plants.

Ammonification is the process by which organic nitrogen is returned to the soil during the decomposition of a dead organism's body.

Nitrification is a way to convert ammonia into nitrate via bacteria. This process sustains plant life because ammonia created during nitrogen fixation is hazardous to plants.

Denitrification turns these nitrates back into nitrogen gas in the atmosphere.

Altogether, these processes work to transfer nitrogen from the atmosphere to plants and the soil and back again. Chemical fertilizers and the cultivation of certain crops, such as soybeans and alfalfa, have caused more nitrogen to be available to live organisms via these processes.

Nitrogen makes the destruction of ozone in our atmosphere occur at a faster rate, and nitrous oxide is one of the three main greenhouse gases contributing to global warming. While nitrogen is vital for life on Earth, it also has the potential to damage life on Earth. It is important to understand these processes and how they can potentially be used to limit the amount of nitrogen in the atmosphere.

Phosphorus

The *phosphorus cycle* moves phosphorus (P) through the lower parts of the atmosphere. Phosphorus is an essential building block of DNA molecules; every living organism requires phosphorous to survive. It is also one of the three major components of commercial fertilizer used in agriculture today.

Phosphorus is not present in appreciable quantities in the atmosphere because it is a solid at ambient temperatures, so it is primarily a terrestrial and lithospheric biogeochemical.

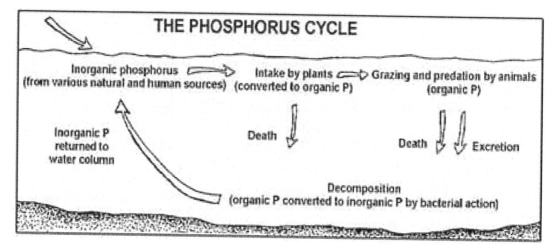

Figure 2.16. *Phosphorus cycle*

Much of the planet's phosphorus is contained in rocks that release phosphate into the soil where plants can absorb it, and it can be eaten by animals. After animals die, the phosphorus in their bodies returns to the soil, and the process continues. It is also excreted daily in phosphates found in animals' manure.

Phosphorus also enters water streams and returns to the ocean, where it can sink and reform in rocks.

Humans are responsible for a significant increase in the amount of phosphorous found in the soil. Fertilizers and animal manure add a great deal of phosphorus to the soil. Since phosphorus is key to plant growth, it makes sense to use it to fertilize crops.

However, a large increase in phosphorous use can overwhelm water ecosystems by a process called eutrophication.

Eutrophication occurs when vegetation overgrows in response to an increase in nutrients and may decrease the level of oxygen in the water available to other organisms. This can suffocate other water organisms and has been a significant pollution problem.

Therefore, understanding the way that phosphorus is cycled through the various biospheres is important in regulating and protecting ecological communities.

Sulfur

Sulfur (S) is an element that is required to sustain life on Earth. It has earned this designation through its presence in two major amino acids used to make proteins, methionine, and cysteine. It is also found in several proteins and polypeptides. Therefore, it is a major component of the proteins that help build a structure in all living cells.

Figure 2.17. *Sulfur formations in White Island, New Zealand*

Sulfur is traditionally found in the Earth's crust via rock or salt domes. Organisms may obtain the sulfur they require from the water runoff from rocks, but that does not necessarily mean that it will be in a form that the organisms can use. That is why the sulfur cycle is so vital to life on Earth; different organisms need to modify sulfur into different forms for it to be used by the organism. These modifications require a series of chemical reactions.

The *sulfur cycle* moves sulfur from minerals into living organisms and back again. Organic sulfur from living organisms is mineralized into compounds, such as hydrogen sulfide, which may be oxidized into sulfates. Plants absorb these, and they enter the food chain. Upon an organism's death, the sulfide contained returns to the ground and is reduced back to sulfate, reentering the sulfur cycle to continue the process.

Sulfur is released into the atmosphere by volcanoes or through sediment runoff and then rains onto the Earth and is absorbed into water and the ground.

Sulfur is released into the atmosphere when fossil fuels are burned, especially coal, which has a high sulfur concentration. Significant increases in atmospheric sulfur have contributed to "acid rain," precipitation that includes sulfuric acid. It is responsible for significant rock and concrete erosion worldwide.

Therefore, it is important to try to maintain the sulfur cycle equilibrium, as too much sulfur in the atmosphere has negative consequences for the entire planet.

Water

The *water cycle* is one of the most important cycles present on the planet Earth, as water is necessary for all life and must be recycled to be available for all living organisms. Water (H_2O) is a major component of all organisms' bodies, a carrier of nutrients, and a major part of the environment. Water starts the cycle by falling as precipitation in various forms (rain, ice, snow, sleet, hail, etc.) onto the Earth, where it is absorbed into lakes, streams, rivers, the ground, and groundwater.

Evaporation from these bodies of water returns the water to the atmosphere. Water then condenses within clouds in the atmosphere to turn into the water droplets that will return to the Earth as precipitation. Air currents move the condensed water in clouds across the planet until the water is condensed enough to become precipitation in one of the previously mentioned forms.

Water can also be released through cracks in the Earth's crust, mainly on the ocean floor, where superheated water jets can erupt. In this way, water is continuously cycling through both the atmosphere and the Earth, so the same water that is present today will be present tomorrow in a different form. It is distributed across the planet in these various forms in a balanced way that allows many different life forms to exist.

Water is not changed during the process; it merely takes different forms during different steps.

This water cycling also contributes to other biogeochemical cycles. It is instrumental in dissolving minerals from rocks into larger bodies of water and supporting those chemicals' cycles as well as their own.

Water is also necessary to cool temperatures on Earth. It takes away heat as it evaporates, so environments warm when there is less water available to evaporate the heat.

The water cycle plays a key role in maintaining life as well as maintaining the status of the world's ecosystems.

Conservation of matter

The law of conservation of matter states that matter is never created nor destroyed in reactions present on Earth. Organic compounds are never created through any cycle but, instead, are transitioned through reactions and matter states.

This means that the same amount of every chemical present at the beginning of Earth continues to be present on Earth; it just may be in a different form or different compound. The ratios all remain the same. They are simply converted into different compounds and types and in some cases, can be converted back by a different process. This conservation of matter allows scientists to conduct experiments and quantify the transfer of compounds across chemical reactions.

Figure 2.18. *Mikhail Lomonosov, a Russian scientist who discovered the law of mass conservation in 1756.*

There are many examples of the law of conservation of matter. A balanced chemical equation is used to demonstrate the equal transfer of molecules across the entire reaction (i.e., starting reactants and the final products).

This means that in every chemical reaction, it is possible to account for all the matter that entered the reaction through what is resolved.

If there are four water molecules at the start of the reaction, there will still be the same number of hydrogen and oxygen atoms at the end of the reaction, even if they are now in different molecules.

This gives people the ability to see the balance in the world's biogeochemical cycles know that they can rebalance these cycles for the betterment of the planet.

Chapter 3

Population

Population Biology Concepts

- Population ecology
- Carrying capacity
- Reproductive strategies
- Survivorship

Human Population: Dynamics

- Historical population sizes
- Distribution
- Fertility rates
- Growth rates and doubling times
- Demographic transition
- Age-structure diagrams

Human Population: Size

- Sustainability
- Case studies
- National policies

Human Population: Impacts of Population Growth

- Hunger
- Disease
- Economic effects
- Resource use
- Habitat destruction

Population Biology Concepts

Population biology is the study of organism populations (group of organisms of a species that interbreed and live in the same place at a same time). Within population biology, the focus is especially given to issues concerning population size regulation, life-history traits (e.g., clutch size), and extinction.

The terms *population biology* and *population ecology* are occasionally used synonymously. When the study of diseases, viruses, and microbes is involved, population biology is the more common choice. When the study of plants and animals is central, population ecology is more appropriate.

Population ecology

Population ecology is the study of the growth, abundance, and distribution of populations. The size of the population is denoted as N (the total number of individuals). Population size in relation to living space is known as *population density* or the number of individuals per given unit of area.

How density is patterned over the population's range is known as *population dispersal.* Populations may be spread uniformly, randomly, or be clumped.

Ecologists study changes in population distribution across space and over time.

Figure 3.1. *This world map indicates the average population density per km².*

Generally, a population is densest near the center of its range and sparsest at the very edge. This edge, the *zone of physiological stress*, has suboptimal conditions for the species in question. Physiological stressors may include extreme temperatures, inadequate water supply, or pollution. Beyond this zone is the *zone of intolerance,* where no individuals of the species can survive. The species' theoretical range is determined by the range of physiological stressors which it can tolerate. However, this range may be restricted by other species due to *biological stressors,* such as competition and predation.

Stressors are *limiting factors*, conditions which act to limit growth or abundance of a population. Limiting factors may be *density-independent,* meaning they are independent of population density; examples include light availability and precipitation. Limiting factors can also be *density-dependent,* whereby the effect becomes more severe as population density increases. Density-dependent factors include competition, disease, parasites, and food scarcity. They typically fluctuate and drive a *population cycle,* a cyclic change in the population size.

The population size over time can be predicted by considering birth rate, or *natality,* and death rate, or *mortality.* Together, natality and mortality can be used to calculate the *intrinsic rate of natural increase (r).*

This is summarized in the equation:

$$r = \frac{(birth\,rate\;-\;death\,rate)}{N}$$

However, the population increase is usually subject to many other factors. For example, population ecologists must often consider the *emigration* of individuals out of the population or the *immigration* of other individuals into the population.

Population growth typically exhibits one of two patterns. The first pattern is *discrete growth,* also known as *discrete breeding* or *discrete reproduction*, an occurrence in which organisms breed all at once at a time of year. They may breed only once in their lifetime, making them *semelparous,* or they may reproduce year after year, making them *iteroparous.*

The former strategy produces *discrete generations* in which the adult generation reproduces and soon dies, leaving behind the next generation. This results in a population that has only one generation at any given time. However, iteroparity produces *overlapping generations* in which an elderly generation is living at the same time as a reproductive generation and a sexually immature generation; at any given time, at least two generations can be observed in the population.

The other pattern of population growth is *continuous growth,* which occurs when organisms reproduce continuously without an established breeding season. Populations that exhibit continuous growth are always iteroparous and have overlapping generations.

Most organisms do not fit neatly into one of these two patterns and instead exhibit a combination of both. For example, plants may reproduce sexually at a certain time each year but also reproduce asexually at any time.

Exponential growth often occurs in populations that are iteroparous and have overlapping generations. It is represented by a J-shaped (*exponential growth*) curve.

The first phase of the curve is the *lag phase*. At this time, growth is slow because the population is small.

At a certain critical size, the population enters the *exponential growth phase*, during which growth accelerates rapidly.

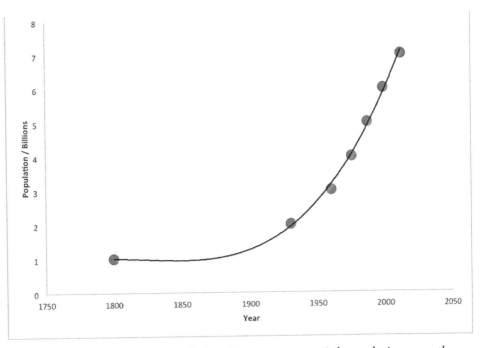

Figure 3.2. *This line graph plots human exponential population growth between 1800 and 2000.*

Carrying capacity

A population experiencing maximum population growth, with no hindrance from limiting factors, is said to be fulfilling its *biotic potential.* The biotic potential is calculated by considering the number of offspring produced by each reproductive event (i.e., *clutch size*), the frequency and a total number of reproductive events, the offspring survival rate, and the age at which an individual reaches sexual maturity.

Populations can only reach their biotic potential when they have ample space, resources, and an absence of predation.

However, these factors are often in limited supply, and the environment can only support so many organisms according to its *carrying capacity* (K).

When populations approach carrying capacity and deplete resources, they encounter *environmental resistance,* and growth begins to slow.

An S-shaped *logistic growth curve* represents growth under environmental resistance. The first portion of the curve is exponential, with the lag phase and exponential phase described earlier.

However, the population eventually reaches a *transitional phase* or *deceleration phase*, during which the birth rate begins to fall below the death rate due to resource competition, predation, disease, and other density-dependent factors. At this point, growth begins to slow.

At carrying capacity, the population enters a *stable equilibrium phase* with minimal growth. During stable equilibrium, natality and mortality rates are roughly equal.

The logistic growth curve can be calculated using the equation:

$$\frac{\Delta N}{\Delta t} = rN \left(\frac{K-N}{K}\right)$$

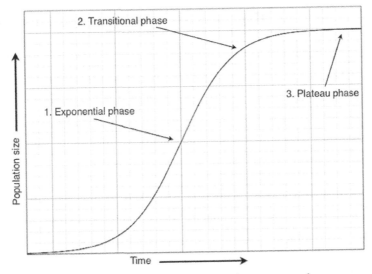

Sigmoid graph showing population growth

Figure 3.3(a). *Graph showing exponential (i.e., unrestricted) growth.*

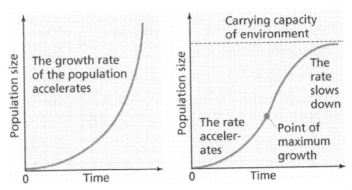

Figure 3.3(b) *Graph showing logistic (i.e., restricted) growth.*

Carrying capacity is determined by the availability of water, space, food, and light, as well as many other factors. It is density-dependent, becoming more restricted as growth increases. Stable populations do not attempt to maximize biotic potential but rather remain just under carrying capacity. However, overshooting carrying capacity can be a valuable strategy as well, provided the population can produce new individuals before it exceeds resource availability.

Carrying capacity is an important population size regulator and a powerful evolution driver. Populations commonly respond to carrying capacity by expanding their range or evolving adaptations, which relieve them of some of the restrictions of carrying capacity. However, no organism can indefinitely evade carrying capacity, and if it is overshot, high mortality will result.

Reproductive strategies

Ecologists have historically divided organisms into two major life-history patterns: *r-selection* and *K-selection*. However, most species are not strictly r-strategists or K-strategists; it is more common for a species to exhibit characteristics of both. They may also be able to shift strategy in response to environmental factors.

Selection

Species that are *r*-selected attempt to maximize their rate of natural increase. They typically overshoot carrying capacity, causing the population to crash suddenly. Therefore, their population growth may show severe fluctuations.

Typically categorized as *opportunistic species*, r-strategists will rapidly seize the opportunity to proliferate. They are often the first to colonize a new habitat and do well in unstable environments subject to density-independent factors.

This is because they reproduce quickly and reach sexual maturity at a young age, maximizing the opportunity for reproduction before they are killed off. They may get the chance only once, making semelparity a common characteristic of r-strategists. The trade-off is that they must produce many offspring at one time because they cannot protect them from infant mortality. To survive, r-selected organisms have a short lifespan and must quickly adapt to new environments.

K-Selection

K-selected species attempt to sustainably maintain their rate of natural increase. They exist near the carrying capacity at a state of equilibrium, making them *equilibrium species*. Unlike r-strategists, K-strategists tend to be specializers uniquely suited to their environment. This makes them very successful but also vulnerable to disturbances. They typically enter a new habitat after r-strategists have already colonized it.

To maintain their existence, K-selected species are typically large and invest considerable energy in caring for their offspring, of which they often produce only one at a time. K-strategists reach sexual maturity very slowly, but they can live for a long time. They can also reproduce several times throughout their lifespan, making them *iteroparous*.

Survivorship

Mortality Patterns

In demography, predictions can be made to ascertain the probability that an individual will die before their upcoming birthday based upon their age. The tool used to make these predictions is a *life table* (also called a *mortality table* or *actuarial table*). This information ultimately signifies the survivorship of certain age-based populations.

In actuarial science, two varieties of life tables are used. In representing mortality rates during set periods for a specified population, a *period table* is utilized. In representing a certain population's overall mortality rates, a *cohort life table* (also called a *generation life table*) is used.

Life tables track cohorts over their lifetime. A *cohort* is a group of individuals born at the same time that are aging alongside one another. The life table shows how members of the cohort begin to die off at various ages.

Levels and trends in the under-five mortality rate, by Millennium Development Goal region, 1990–2012 (deaths per 1,000 live births, unless otherwise indicated)

Region	1990	1995	2000	2005	2010	2012	MDG target 2015	Decline (percent) 1990–2012	Annual rate of reduction (percent) 1990–2012	1990–2000	2000–2012
Developed regions	15	11	10	8	7	6	5	57	3.8	3.9	3.8
Developing regions	99	93	83	69	57	53	33	47	2.9	1.8	3.8
Northern Africa	73	57	43	31	24	22	24	69	5.4	5.3	5.5
Sub-Saharan Africa	177	170	155	130	106	98	59	45	2.7	1.4	3.8
Latin America and the Caribbean	54	43	32	25	23	19	18	65	4.7	5.1	4.4
Caucasus and Central Asia	73	73	62	49	39	36	24	50	3.2	1.6	4.5
Eastern Asia	53	46	37	24	16	14	18	74	6.1	3.7	8.0
Excluding China	27	33	31	20	17	15	9	45	2.7	–1.2	5.9
Southern Asia	126	109	92	76	63	58	42	54	3.5	3.1	3.9
Excluding India	125	109	93	78	66	61	42	51	3.3	3.0	3.5
South-eastern Asia	71	58	48	38	33	30	24	57	3.9	3.9	3.8
Western Asia	65	54	42	34	26	25	22	62	4.4	4.4	4.5
Oceania	74	70	67	64	58	55	25	26	1.4	1.0	1.7
World	90	85	75	63	52	48	30	47	2.9	1.7	3.8

Note: All calculations are based on unrounded numbers.

Figure 3.4. *This life table, developed by the United Nations, indicates the mortality levels and trends for children under five years old worldwide.*

Survivorship refers to how many individuals remain alive at a given point.

The three general types of survivorship curves (shown below) are:

Type I survivorship curve, which shows a long curve with a relatively short drop-off near the end in which most individuals survive until they die of old age (such as in the human population).

Type II survivorship curve, which is negative and linear, shows individuals dying at a constant rate over their theoretical lifespan (as is the case with some bird and lizard species).

Type III survivorship curve is seen when most individuals in the population die at an early age. Those that do survive, however, tend to live for a relatively long time. It is essentially the opposite of a Type I curve and is seen in many invertebrates, plants, and fish.

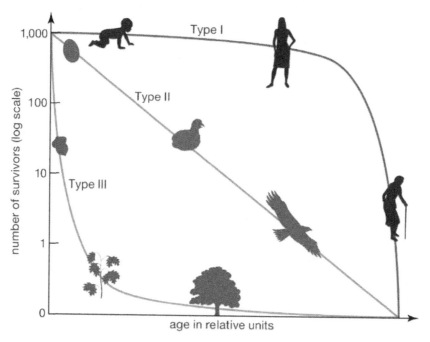

Figure 3.5. *Survivorship curves*

Human Population: Dynamics

Historical population sizes

The human population is now in the exponential phase of a J-shaped growth curve. The global population is currently increasing by about 80 million people per year. This tremendous growth has been fueled by technological advances, an increase in the food supply, reductions in disease, and habitat expansion.

The world population signifies the total number of people currently alive, and 7.4 billion was the tally made in March 2016.

Current estimates predict that the world population will level off sometime this century at 8-10 billion people. Some claim this number is a vast underestimation and that the human population may reach 14 billion by the end of the 21st century.

In the mid-20th century, the most developed countries had a significant decline in mortality rate, soon followed by a decline in the birth rate. Currently, the growth rate in first world nations is at about 0.1%. Their age structures are relatively stable, with some countries even exhibiting a declining population.

Nearly all population growth in the coming years will be seen in the less developed countries, especially in Africa, Asia, and Latin America. Even so, the growth rate in these countries is still lower than it was at its peak of 2.5% in the 1960s.

The only way to level off population growth is an aggressive campaign of family planning, birth control, and encouragement to produce fewer children.

However, due to cultural attitudes and a high infant mortality rate in developing nations, it is difficult to convince people to have fewer children or delay childbearing until later years.

Figure 3.6. *London, with approximately 6.5 million residents, was the most populous city in the world at the turn of the 20th century.*

Distribution

Both the growing populations of the less developed countries and the high consumption found among more developed countries put stress on the environment. Currently, the *ecological footprint* of those in the United States and other first-world nations is unsustainable. The ecological footprint is determined by the amount of land required to sustain an individual's lifestyle, including the area in which they live, the farmland required to produce food, the factories required to produce material goods, and the distances across which these products must travel to reach the individual.

An average American family, in terms of consumption and waste production, is comparable to the lifestyle habits of thirty average people in India. Developed countries account for one-fourth of the world population but provide 90% of the hazardous waste production.

Intense resource consumption affects the cycling of chemicals and contributes to pollution and the extinction of species. Without drastic reductions in the collective ecological footprint, humans will soon overshoot carrying capacity and possibly experience catastrophic disease, famine, and other density-dependent catastrophes.

Fertility rates

The average number of children a woman has in her lifetime in a certain population is a *total fertility rate* (TFR). It is also referred to as *fertility rate*, *period total fertility rate* (PTFR), or *total period fertility rate* (TPFR). This rate predicts the average number of children for a woman if:

1) She experienced the precise, current age-specific fertility rates through her lifetime; and

2) She survived all childbirths during her reproductive life.

To find the total fertility rate, the single-year, age-specific rates at a given time are summed up.

Not being based on any real group of women's fertility rate (since such a measurement would involve waiting until they had completed childbearing) TFR is a synthetic rate. TFR is also not based on summing the group of women's total lifetime number of children born. The age-specific fertility rates of women in the 15–44 or 15–49 age range is what TFR is based upon instead. These two age ranges reflect conventional international statistical usage.

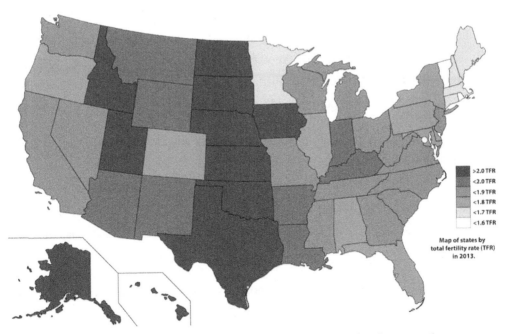

Figure 3.7. *This United States map indicates the 2013 total fertility rates by state.*

In sum, TFR is a demographic measurement centered around an imaginary, everyday woman. This woman, as she passes (and survives) through her reproductive years, is subject to all age 15-49 fertility rates that were recorded for a given population in a particular year. So, under the premise that this woman can be seen (in a fast-forward manner) passing through her reproductive years, subject to all the age-specific fertility rates calculated in a specific year, the resultant TFR represents the average number of children this woman would have.

Net reproduction rate (NRR) is an alternative fertility measurement. NRR, operating under the premise that a woman is subject to a given year's age-specific fertility and mortality rates, calculates the total number of daughters she would give birth to during her lifetime. While the use of the NRR has been phased out in many contexts (such as the United Nations since 1998), it is still relevant and in use for countries that experience high gender imbalance ratios (e.g., China and India). Like NRR is the *gross reproduction rate* (GRR), which unlike the NNR ignores life expectancy.

A more accurate index of fertility rates is obtained using the TFR or TPFR, as opposed to relying only on the crude birth rate (annual number of births per thousand population). This accuracy stems from the TFR and TPFR's independence from the population's age structure. This independence, however, does result in the negative effect that it provides a less clear estimate of actual completed family size. For that goal, the total cohort fertility rate is preferred since it signifies the sum of the applicable age-specific fertility rates to each cohort. Another deficiency of the TFR is that it does not account for generational differences in birth rates (e.g., women of younger generations having fewer children than women from older generations).

The minimum fertility rate needed to precisely sustain (i.e., zero growth rate) a certain population level is the *replacement fertility rate*. Hypothetically, if there were no childbirth deaths of mothers during their reproductive years (which is highly improbable), the resultant replacement fertility rate would be exactly 2.0 (i.e., two children per woman). In the U.K., an industrialized nation where people generally have easy access to enough healthcare services, the replacement fertility rate is 2.075.

Developing nations with higher mortality rates, however, have replacement fertility rates that typically range between 2.5 and 3.3. Globally speaking, the replacement fertility rate is 2.33. Interestingly, an alternative view of the relationship between a population's economy and its reproduction rate is provided by a log-transformation of the data, which reveals that for every 1% *Gross Domestic Product* (GDP) gained, the TFR is reduced by 0.26%.

Growth rates and doubling times

The total number of live births (not including stillborn births) per 1,000 people in a year is the *birth rate*. Typically, census data or registration systems for births, deaths, and marriages are utilized to calculate a birth rate.

Used in combination with mortality and migration rates (both immigration and emigration), the birth rate can be used to calculate population growth.

The rate that signifies the increase in the number of individuals in a population for a specified period is the *population growth rate*, which can be expressed as both a fraction and a percentage. The formula used to calculate this rate is shown below:

$$Population\ growth\ rate = \frac{P(t_2) - P(t_1)}{P(t_1)(t_2 - t_1)}$$

Rates showing a population gain indicate a positive growth rate; rates showing a reduction in population indicate a negative growth rate. A population growth rate of (near) zero, indicating neither a gain nor a loss in a total number of individuals, might be challenging to understand when changes in birth, death and migration rates, as well as the age distribution over time, are unaccounted for. Likewise, wide-scale loss of life due to disease and conflict can affect population growth rates.

The *net reproduction rate* is a measurement premised on the absence of migration. In interpreting the meaning of the net reproduction rate, a value greater than 1 signifies an increase in the female population, while a value of less than 1 signifies a decrease in the female population (sub-replacement fertility).

The demographic condition in which the number of individuals neither increases nor decreases is *zero population growth* (ZPG, also called *replacement level of fertility*).

This figure includes differences caused by migration, emigration, births, and deaths. According to certain schools of thought, ZPG is a desirable state to attain in efforts to achieve long-term environmental sustainability considering contemporary concerns about overpopulation.

The amount of time that must pass for a stated figure to double is called the *doubling time*.

Doubling time's key properties are:

1. Larger rates of growth correspond with quicker doubling times;

2. Growth rates for organisms vary, especially when the size of the organism is considered. Generally, smaller organisms (e.g., bacteria) both grow faster and reproduce at a quicker rate than larger organisms (e.g., elephants); and

3. There is usually a limit to population size. Natural resource constraints and disease contribute to reigning in the size of populations. The growth that accounts for the influence of resource availability and disease is called *logistic growth*.

Doubling time is evident in the human population. The increase from 500 million people to one billion people in 1804 took approximately 300 years. Subsequently, the doubling time that it took to get up to 2 billion people in 1927 was approximately 123 years, and from then, it took an additional 47 years to reach four billion people in 1974.

By approximately 2025, the human population is expected to reach eight billion people, according to United Nations' projections.

When a human population, without experiencing either high mortality rates or high emigration rates, sustains a 3.8 TFR for a protracted period, its doubling time will be approximately 32 years. Populations maintaining zero growth rate and not subject to mass immigration will experience an overall population drop over time.

Since it will take several generations for equilibrium to be established in the age distribution, it will take a similar amount of time for changes to appear in the total fertility rate. Even when abruptly dropping below the replacement-level fertility rate, for example, a population will continue to grow due to many young couples (produced by the previously high fertility rate) now entering their reproductive years.

Carried for several generations, this would result in a *population momentum, population inertia* or *population-lag effect*. Population inertia is a population's tendency to maintain its current density by resisting changes.

Population momentum refers to the phenomenon where national population levels grow even if childbearing levels declined to replacement level. Population-lag effect refers to the rapid population growth a nation experiences when a total fertility rate of at least 3.8 is sustained for an extended time without an associated rise in death or emigration rates.

When growth increases in a constantly proportional manner, this is called *exponential growth*, which can result in many environmental strains on society. Due to Earth's finite resources, exponential growth has resulted in climate change, biodiversity loss, and deforestation.

Demographic transition

As a nation transforms from pre-industrialization to an industrialized economic system, the commonly resulting transition from high birth and death rates to low birth and death rates is a *demographic transition* (DT), which can be demonstrated by a corresponding *demographic transition model* (DTM).

American demographer Warren Thompson (1887–1973) provided the basis for this theory through his 1929 interpretation of demographic history, having noted birth and death rate changes over the previous 200 years in industrialized countries.

The demographic transition is composed of four (possibly five) stages:

Stage 1: Before the late 18th century in Western Europe, as indicated by research conducted on the history of human populations, all pre-industrialized nations had high death rates and high birth rates — nearly a state of equilibrium.

Between the First Agricultural Revolution of 10,000 B.C.E. and the onset of industrialization in the mid-19th century, growth rates were at 0.05% or below, signifying very slow growth rates due in large part to a lack of the food production required to sustain more rapid reproduction.

Stage 2: An industrializing nation, benefitting from the resultant improvements in food supply (e.g., selective breeding practices, crop rotation), sanitation, medicine, prolonged lifespans, access to technology, and expanded access to healthcare and education, typically experience a rapid decline in death rates.

When numerous fundamental improvements in public health are made (e.g., food handling, water supply, sewage, personal hygiene), mortality rates (and especially childhood mortality rates) notably decrease. This trend could result in a large population increase if a drop in birth rates does not accompany it.

Stage 3: Access to contraception, wage gains, urbanization, the decline of subsistence agriculture, women's gains in both social status and education access, the disappearance of child labor, the bolstering of child education, and other fertility factors cause birth rates to fall. A leveling off in population growth manifests, as evidenced by the 19th-century trends in Northern Europe.

Stage 4: Populations experience low birth rates and low death rates. A dwindling population becomes the result if birth rates plummet too far under the replacement level, becoming a threat to industries that rely on positive population growth.

The aging of the population born during industrialization (stage 2) and their associated care needs to cause a significant economic burden for the shrinking population (e.g., contemporary Japan's aging population situation).

Negative behaviors and lifestyle choices, such as lack of physical activity and overeating, may lead to a rise in disease. Birth and death rates leveled off in developed nations by the late 20th century.

Demographic transition models, despite contradictory data in recent history indicating fertility rates resume rising after a certain development point, continue to predict decreasing fertility rates.

Additionally, despite the establishment of a matching demographic transition correlation, it remains uncertain whether diminished populations are the result of industrialization and higher incomes or vice versa.

While some exceedingly impoverished nations in Africa and Asia have yet to reach full industrialization levels sufficiently, most nations have at least reached stages 2 or 3. Many world populations have already attained stages 3 or 4.

Age-structure diagrams

A population with overlapping generations typically exhibits *pre-reproductive,* *reproductive,* and *post-reproductive* generations. The abundance of individuals of each gender and in each age group of a population is represented in an *age structure diagram.*

Horizontal bars represent the number of individuals in each age group.

A pyramid shape indicates an expanding population with a high birth rate and exponential growth. The pre-reproductive generation is largest because offspring are being rapidly reproduced, while the reproductive generation is intermediate.

The post-reproductive generation is smallest as the elderly die off.

A bell shape represents a relatively stable population, in which the pre-reproductive and reproductive generations are roughly equal, and the post-reproductive generation is smaller by only a narrow margin.

Finally, an urn-shaped diagram is indicative of a declining population. The post-reproductive generation is largest because few new individuals are being produced.

Individuals from the reproductive generation rapidly enter the post-reproductive generation and continually die off, while the pre-reproductive generation is too small to sustain growth.

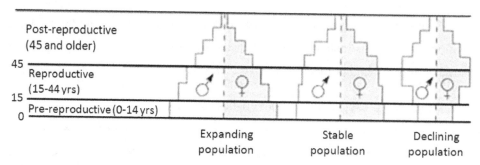

Figure 3.8. *Age structure diagrams: pyramid for expanding (left), bell-shaped for stable (center), urn-shaped for a declining population (right).*

Human Population: Size

Sustainability

The world population, according to United Nations' statistical data, grew by 30% between 1990 and 2010, resulting in approximately 1.6 billion more people. India was the nation that grew the most rapidly (at a 350 million increase) with China in second place (with an increase of 196 million). According to percentages of population increase, the United Arab Emirates (315%) and Qatar (271%) experienced the greatest positive change.

A powerful teaching tool – the concept of exponential population growth – presents its potential when consideration is given to how it affects environmental system health. The implications are stark; the greater the human population on Earth, the greater the strain it puts on the environment. Population growth and economic growth are cited by the *United Nations Population Fund* (UNFPA) as two of the major problems facing environmental quality.

Population growth and its negative implications can be addressed in several ways. This importantly includes promoting voluntary family planning (e.g., offering sex education, ease of access to birth control and contraceptives) and ensuring basic rights (e.g., healthcare, education, and economic opportunity) to women. These approaches help mitigate the burden of overpopulation via empowerment, not coercion.

Case studies

The fastest-growing populations are typically those located in *less economically developed countries* (LEDC). In the demographic transition model, most of these countries have reached stages 2 or 3, characterized by birth rates increasingly outpacing death rates and resulting in a population boom.

Death rates are declining due to improvements in healthcare and sanitation.

Several reasons help explain the high birth rates:

1. Inadequate access to contraceptives and family planning education.

2. Rural and urban demand for child labor in the agricultural and informal sectors, which is often deemed necessary by families struggling to make ends meet.

3. High levels of infant mortality, leading women to habitually have more children.

4. Religious and cultural norms that forbid the use of contraceptives.

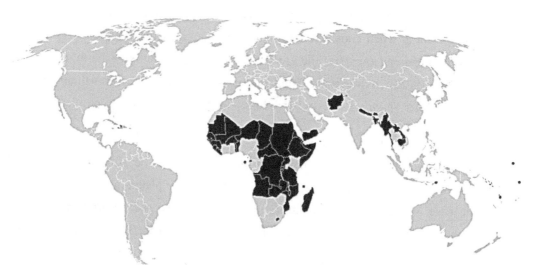

Figure 3.9. *This ECOSOC world map highlighting the current LEDCs.*

With a population of approximately one billion people, India currently possesses one of the highest growth rates worldwide. For example, 181 million people (approximately the combined populations of neighboring Bangladesh and Sri Lanka) were added to India's population in the last ten years alone.

If this growth rate is sustained, by 2020, India will surpass China as the most populous nation in the world.

This growth rate is slowing, particularly in India's southern state of Kerala.

There, initiatives to reduce population growth have been implemented, including:

1. The advances in women's education that has seen Kerala's female literacy rates rise to 85% (compared to a national average of ~50%), helping to lower infant mortality rates due to mothers' increased ability to care for them, which in turn has resulted in lower birth rates (since couples now tend not to have more children in anticipation of compensating for those who might die young);

2. Increased access to contraceptives; and

3. The improvement of women's status in a society where they are increasingly viewed as an asset, not a liability or burden (as is evidenced in Kerala's shifting marriage custom where the bridegroom's family increasingly pays the dowry instead of the bride's family).

In China, the 1979 implementation of the "One Child Policy" aimed to curb the nation's growth rate by incentivizing families with free education, healthcare, pensions, and other benefits if they limited themselves to having only a single child.

All births in China under this system needed prior approval from family planning officials. If a family ignored the policy and had additional children beyond the allowed, their benefits were revoked, and penalties were sometimes levied.

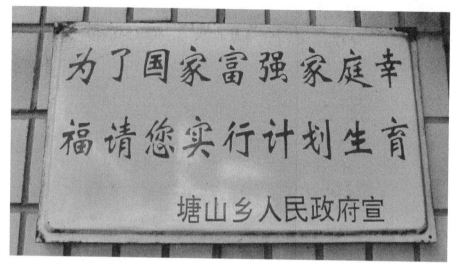

Figure 3.10. *This government sign in Nanchang roughly translates to: "Please, for the sake of your country, use birth control."*

Under the implementation of China's One-Child Policy, its population growth rate declined, preventing an additional 320 million children being born (about the size of the United States). This policy was not without its negative or problematic consequences.

Since traditional Chinese Confucian beliefs often result in male children being more valued than females (to carry on the family name), female infant and child abandonment became a serious issue. Additionally, due in large part to this policy, approximately 90% of the fetuses aborted during the height of the One-Child Policy were female.

More economically developed countries (MEDCs), on the other hand, typically have stabilized population growth with birth and death rates. Problematic situations, however, are not absent from MEDCs.

For example, in Germany, the death rate is higher than the birth rate, meaning that their population is steadily declining at about –0.1%. This contributes to an increasingly aging population, and when combined with the reality of increased longevity due to healthcare, diet, and lifestyle improvements, the German nation faces a future filled with an increased number of elderly dependents. The U.K., facing a similar situation, seems likely to address the financial demands of an aging population by increasing taxes to fund pensions and healthcare needs.

National policies

Policies that respect parents' reproductive goals, promote women's health and are sensitive in supporting an educationally and economically active society are evidenced around the world.

However, many nations and territories continue to halt this progress by cultural resistance and political infeasibility, even despite the reality that these implemented policies have proven to be inexpensive.

Aligning the human population with an environmentally sustainable trajectory can be accomplished by several proposed strategies:

- *The provision of both sexes with universal access to safe and effective contraceptive options.* Redressing the lack of adequate family planning services and access, as evidenced by the ratio of two out of five pregnancies being reported as unplanned, could close the assurance gap that each child conceived is wanted and welcomed.

- *The guarantee of secondary education for all children, including girls.* Women who have attained some level of secondary school education, in all surveyed cultures, tend to have fewer children in general, and those they do have are born later in the mother's life compared to the reproductive behaviors of less-educated women.

- *The eradication of biases of gender in law, economic opportunity, health, and culture.* Postponing childbearing and limiting the number of children is more commonplace among women who are not deprived of the rights to own, inherit and manage property, obtain credit and participate in civic and political affairs as are men.

- *The universal provision of age-appropriate sex education.* The exposure to comprehensive programs on sex, sexual development, sexuality (sexual orientation), and reproductive health, as indicated in data gained in the United States, can ameliorate the rate of unwanted pregnancies and lowering birth rates.

- *The elimination of child-birth incentives.* Taxes (or other financial benefits) can be preserved or even increased by governments in aims to assist parents in ways that are based on the status of parenthood itself, as opposed to incentives based on the number of children born to a family.

- *The integration of population, environment and development lessons at multiple levels into school curricula.* It is imperative for schools to provide knowledge for students to make well-informed decisions (e.g., to have children or not; and when) on how to navigate and manage their comprehensive effect on the environment.

- *The pricing of environmental costs and impacts.* Couples cognizant of the financial ramifications of childbearing can forecast the incentives and disincentives attached to the size of a family.

  Calculating taxes, the cost of food, clothing, and other associated costs incurred per child, as well as having an awareness of available government rebates, could impact the number of children couples would desire.

- *The refrain from encouraging childbirth through government incentives and programs, instead of adjusting to an aging population.* Societal adjustments (e.g., increased labor participation) must be implemented to manage an aging population responsibly.

- *Promoting leaders to adopt policies to stabilize population growth via human rights and human development.* Policymakers can address problems associated with population ethically and effectively by educating themselves on rights-based population policies and encouraging women to make informed reproductive choices through empowerment.

Human Population: Impacts of Population Growth

Planetary boundaries (definitions for responsible societal and governmental stewardship of Earth to ensure its continuing habitability) were said to have been crossed by human activity in four critical areas in a study published in *Science* in January 2015.

The four currently transgressed planetary boundaries are climate change (e.g., CO_2 atmosphere concentrations), biosphere integrity loss (i.e., mass extinction of species), land-system change (e.g., deforestation) and biogeochemical cycles (e.g., phosphorus and nitrogen flows).

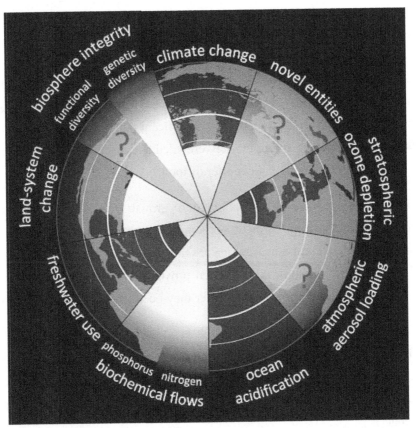

Figure 3.11. *This planetary boundary diagram indicates the variety and current integrity of the boundaries. The length of the white wedges radiating out from the center indicates the level of danger associated with human behavior. The closer the silver wedge is to the center, the greater the safety margin; the further from the periphery, the greater the danger.*

The relationship between population and climate change warrants closer examination. Since the Industrial Revolution (and its associated boom in population size), carbon dioxide emissions have significantly increased.

Fast-forward to more modern times, CO_2 emissions in 2011 were 150 times worse than levels in 1850, indicating a drastic alteration in the atmosphere's composition.

By 2014, the highest levels of CO_2 concentration found on Earth in millions of years were recorded at 400 parts per million (PPM). A graphing of population growth and CO_2 emissions over time reveals a strong positive correlation, indicating that inextricable links exist between population dynamics, consumption patterns, and climate change.

Hunger

In assessing the world's food situation, there is no noticed correlation between hunger and population density. In Bangladesh, a densely populated nation, hunger exists. In Nigeria, Brazil, and Bolivia, nations with significant per capita food resources, hunger exists. In the Netherlands, a small, densely crowded nation, hunger has been eliminated, and they have become a significant exporter of food.

In examining the world's population growth trends, there is an evident correlation between the rapid growth of population and hunger. The fastest-growing nations in Asia, Africa, and Latin America are those that also experience the lack of food resources. However, causation has not been proven. Understanding whether hunger is caused by rapid population growth or whether they occur simultaneously due to the same societal realities is important.

Remarkable insight was gained through Cornell University's 1989 study of 93 developing countries' population growth, food consumption, and other variables. No evidence in their statistical analysis suggests that the rapid growth of the population causes hunger.

Instead, what they discovered was that people had less to eat in poorer countries, particularly those in which the poorest 20% of people earned the smallest share of their nation's total income. In other words, poverty and inequality, not the rate of population growth, cause hunger.

The 11.3% population rise between 2000 and 2009, which brought Earth's number of people from 6.085 to 6.775 billion, was not one of uniform growth. Developing nations (with low per capita income) showed the highest rates; these were largely the same nations with the most severe hunger and malnutrition.

Considering this data, one might think that childhood malnutrition increased at a similar rate as the world population. However, that is not the case. World Bank data reports that for children under the age of five, the rate of malnutrition decreased by more than 3% (24.61% in 2000, 21.27% in 2009), showing that even as the world's population has grown, the proportion of malnourished children declined. This leads to the conclusion that there is no strong relationship between overpopulation and the world's ability to fight hunger.

Disease

Humanity will likely change the way disease epidemic prevention and treatment are handled due to increases in international trade and travel, the sheer population growth, and humans' ever-increasing interactions with animals and ecosystems.

The 2.5 billion to 6 billion population growth in the second half of the last century has likely influenced how infectious diseases emerge and spread.

The rate of new pathogen-caused emergent diseases, when analyzing outbreaks from the mid-20th century up till now, has increased notably over time. This is the case even when controlling for improvements in diagnostic techniques and monitoring. This could result in the false perception that diseases were on the rise. A study found that between 1940 and 2004, more than 300 new infectious diseases emerged.

Pathogens hopping from one species to the next that eventually infected humans accounted for some of these new diseases. Examples of these transmission varieties are the West Nile virus, the SARS coronavirus, and HIV. Others are new drug-resistant variants of existing pathogens (e.g., tuberculosis and malaria).

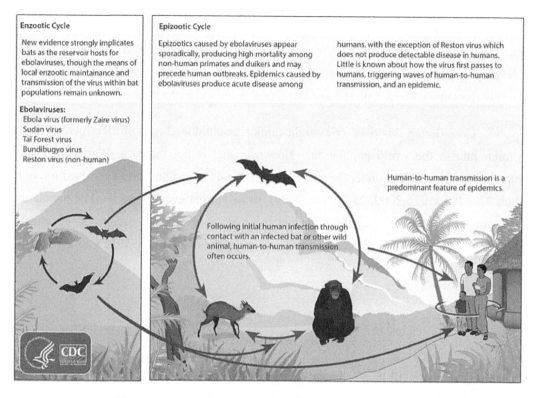

Figure 3.12. *Diagram showing the Ebola animal-animal and animal-human transmission cycle.*

Also, on the rise are occurrences of certain diseases, such as the bacterial-caused (and deer-carried) Lyme disease, which suggests the causation is changed in human patterns of interaction with environments and the disease-carrying animals that inhabit them.

Some parts of Earth (such as tropical Africa, Latin America, and Asia) are also more likely to produce new viruses; experts refer to them as *hotspots*.

A rise in human interactions with environments supporting high biodiversity is a likely cause for viral transmissions from animals. Given the globalization of modern times, these illnesses can quickly spread to anywhere on the globe.

Sea and ground travel that previously would have required months can now be accomplished in a few hours by flight. While this has been beneficial on many levels, it does promote the spread of diseases to and from far-flung places.

For example, an asymptomatic passenger from San Antonio, Texas, arriving in Shanghai can easily result in the introduction of new pathogens to that region. Adding the reality that Earth's projected population growth will result in more global travelers, it is likely that contagions will spread more quickly, contributing to worse epidemics.

In response to humanity's continued growth, health authorities are advocating for both more robust public health organizations and increasing the distribution of resources to protective systems that serve the public.

For vaccine development to be expedited, researchers are exploring methods for quickening viral identification. Scientists are also attempting to gain deeper insights into the complicated series of human-ecosystem interactions to identify emergent disease hotspots. It is hoped that these efforts will be able to ameliorate pandemics by introducing new and creative solutions.

Economic effects

Colloquial use of the term "overpopulation" suggests that full employment and guaranteed access to a decent standard of living might be unattainable due to more people in the economy than can be supported, but a more nuanced understanding is that an economy's capability to employ is dependent on a base of capital needs requiring people for its operation. Capital assets are lacking in overpopulated economies.

Also, one method to increase the rate of capital asset acquisition – installing new industries in the economy – is hindered because the associated savings rate is low. Such an economy risks lowering per capita income and reducing the savings percentage of people since the total production is consumed by more people than are required to generate it.

The need for more goods to sustain a population is also defined by overpopulation, a state in which there are not enough resources (e.g., goods, supplies, water) to share per person.

Overpopulation also strains education systems since a greater number of pupils can lead to a reduction in a school's available funding. Classroom overcrowding and the associated decrease in one-on-one teacher-student time foster a reduction in education quality. Like classroom overcrowding, overpopulation can negatively affect people's access to housing.

Figure 3.13. *Thirty-two children in a small classroom near Nagar, Pakistan.*

Resource use

The outpacing growth of population compared to the availability of both critically needed non-renewable and renewable resources over the past several decades means that despite technological advances, resources per person are in decline. Declining resources are evidenced by the extinction of species, the destruction of rainforests, desertification, the advance of urban sprawl, growing water shortages, toxic waste, oil spills, and air and water pollution.

Finite resources (e.g., fossil fuels, fresh water, arable land, coral reefs, frontier forests) continue to diminish as humanity breaks new population records every year. This encourages an increasing level of competition for vital resources, diminishing the base quality of life.

Involving roughly 1,400 scientists and five years of efforts, a *United Nations Environment Program* (UNEP) Global Environment Outlook study discovered the consumption rates of humanity far outweighed resource availability. Scaling this down to a micro perspective, to meet their various resource needs, each currently in existence requires one-third more land than the planet can supply.

The four-year *Millennium Ecosystem Assessment* research effort, involving 1,360 leading scientists to measure the valuable natural resources have to humanity and the planet, concluded that "the structure of the world's ecosystems changed more rapidly in the second half of the twentieth century than at any time in recorded human history, and virtually all of Earth's ecosystems have now been significantly transformed through human actions."

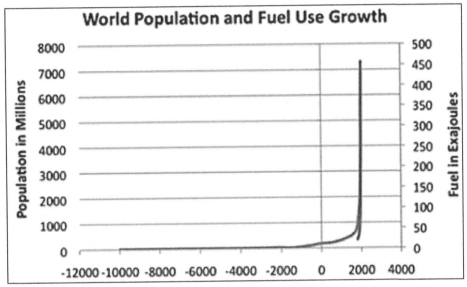

Figure 3.14. *World population from U.S. Census Bureau overlaid with fossil fuel use.*

In research conducted by the Global Outlook for Water Resources to the Year 2025, it is concluded that more than half of Earth's population will become water-vulnerable by 2025 and that water demand will account for 70% of the available freshwater. Another report, one prepared by the 2030 Water Resources Group in 2009, claims that by 2030, some developing areas worldwide will experience water demand that will exceed the supply by 50%.

By 2030, a United Nations report commissioned by more than two dozen of its bodies claims that acute water shortages will affect nearly half of the world's population. Reports such as these provide the basis for the U.N.'s proclamation that Earth is experiencing a "Global Water Crisis." In North Africa, the Middle East, the Indian subcontinent, China and the United States, freshwater is being consumed ten times faster than the rate of replenishment despite it's vital, finite and irreplaceable nature.

The diverse populations located in lakes, rivers, and wetlands (characterized as freshwater ecosystems that account for about 1% of the Earth's surface) could live in some of the most endangered ecosystems on the planet. Research by the *International Union for Conservation of Nature* (IUCN; formerly the World Conservation Union), an organization that tracks biodiversity threats, claims that 34% of freshwater fish species currently risk extinction.

As can be seen by the loss of a larger proportion of species and habitat than any other land or ocean-based ecosystems, freshwater ecosystems are particularly vulnerable to irreparable damage, species endangerment and extinction due to river damming, water pollution, overfishing, and other dangers.

Habitat destruction

Extensive loss of rainforests, coral reefs, wetlands, and Arctic ice, among other ecosystems, can be largely attributed to human overpopulation. Once covering roughly 14% of the planet's land area, the rainforest has been reduced to 6% and could disappear in the next 40 to 84 years if deforestation rates continue at their current trajectory.

Likewise, coral reefs, due to warming temperatures and increasing ocean acidity and pollution, could be decimated by the century's end. By 1980, nearly 30% of reefs disappeared. Included in this figure is the 90% destruction of the Philippine reefs, the 50% destruction of the Caribbean reefs.

Another aspect to consider is 11.5% per decade rate of permanent ice cover loss relative to averages from 1979 to 2000. In 4 to 30 years this sustained rate of loss could result in ice-free Arctic summers. Humanity's footprint also reduces wetlands worldwide on interconnected, ecological infrastructures.

In the United States, about 47% of the original wetlands remain, which is a loss of about 104 million acres. In Europe, 30-40% of wetlands remain. As the world's population continues to grow, so will its footprint on the ecosystems.

The cause of Earth's greatest mass loss of species since the extinction of dinosaurs 65 million years ago is humanity. This humanmade extinction rate is 1,000 to 10,000 times faster than normal. The *International Union for Conservation of Nature's*

(IUCN's) *Red List of Threatened Species* (2012 update) indicates that 19,817 of the 63,837 species examined (nearly a third) are threatened with extinction.

Scientists caution that if present trends continue, at least 50% of the Earth's plant and animal species will become extinct within the next few decades. The causes identified for this are acidifying oceans, climate change, habitat loss, human overpopulation, invasive species, natural resource exploitation, overfishing, and poaching.

In assessing other facts (such as the human population's annual absorption of 42% of terrestrial net primary productivity, 30% of marine net primary productivity, 50% of fresh water, 40% of land allocated to food production and 50% of land developed) it is evident that human overpopulation continues to dominate Earth's physical, chemical, and biological conditions and limits.

Extinction records indicate that the average (not humanmade) extinction rate was one species lost per million species per year. Compare this with the present-day loss of 30,000 species per year (or three per hour), indicating a rate that outpaces nature's rate of adding new species through evolution.

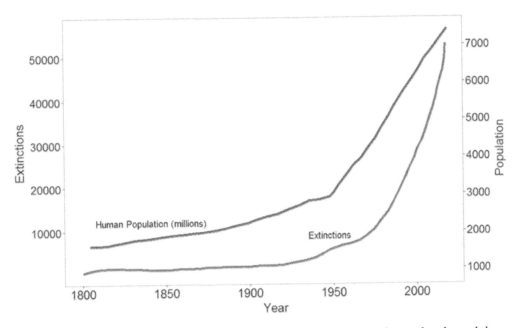

Figure 3.15. *Chart showing the correlation between human population levels and the number of species lost to extinction.*

Notes

Chapter 4

Land and Water Use

Agriculture
- Feeding a growing population
- Controlling pests

Forestry
- Tree plantations
- Old growth forests
- Forest fires
- Forest management
- National forests

Rangelands
- Overgrazing
- Deforestation
- Desertification
- Rangeland management
- Federal rangelands

Other Land Use
- Urban land development
- Transportation infrastructure
- Public and federal lands
- Land conservation options
- Sustainable land-use strategies

Mining
- Mineral formation
- Extraction
- Global reserves
- Relevant laws and treaties

Fishing
- Fishing techniques
- Overfishing
- Aquaculture
- Relevant laws and treaties

Global Economics
- Globalization
- World Bank
- Tragedy of the Commons
- Relevant laws and treaties

Agriculture

Agriculture is a major industry in the United States. The U.S. has been a leader in seed improvement, as evidenced through the creation of bioplastics and biofuels, which would have been impossible without the groundbreaking work of George Washington Carver. Members of the U.S. agricultural sector have made great contributions with the invention of such revolutionary products as John Deere's steel plow, Cyrus McCormick's mechanical reaper, Eli Whitney's cotton gin, and the Fordson tractor.

Figure 4.1. *Cranberry harvest, New Jersey*

Feeding a growing population

The world's population is expanding rapidly, and as a result, there is an ever greater need to devote resources to food security so that every human being has the food they need to thrive.

The greatest challenge is producing nutritious food that is accessible to everyone. If current estimates are correct, Earth's population will be 9.2 billion by 2050. This means that food production must increase by 70% to meet global needs. Steps are being taken to resolve this problem.

One method is the creation of a sustainable food production system, where methods of growing, processing, and disposing of food are undertaken to yield the largest possible return while expending the fewest resources. This can also involve the utilization of renewable resources like recycled waste for fertilizer.

Human nutritional requirements

The basic nutritional requirements of humans include carbohydrates, proteins, fats, vitamins, minerals, and water.

Carbohydrates include starch and dietary fibers, which provide the calories needed to fuel the muscles and the brain. Carbohydrates are found in insoluble fiber, whole grain bread, vegetables, fruit skins, etc.

Protein is found in meat, fish, milk, eggs, and green vegetables, and it is broken down into amino acids by the digestive system. Amino acids make hormones and repair muscles, red blood cells, hair, and other tissues.

Fat, which can be either saturated or unsaturated, generally maintain hair and skin and provides insulation.

Vitamins help regulate chemical reactions in the human body.

Minerals, such as calcium and magnesium, aid in the creation and maintenance of healthy bone and tissue structure and participate in other countless bodily processes. The mineral iron helps red blood cells transport oxygen.

Finally, *clean water* is vital for maintaining life on all levels. The human body is 70% water, which helps control body temperature, carries vitamins and minerals to cells, and eliminates waste.

Types of agriculture

There are many types of agriculture practiced around the world. The following are the essential agriculture types:

Nomadic herding: the people of semi-arid and arid regions mainly practice this type of agriculture, which is based on herding animals into pastures. To find more nutritious and less exploited pastures for their animals to graze, those who practice nomadic herding move from place to place.

Livestock ranching: this type of agriculture, as opposed to nomadic herding, is sedentary and develops a commercial base in areas containing many grazing animals.

Shifting cultivation: this type of agriculture is practiced in rainy, tropical areas and uses *slash and burn* techniques to create new plots for farming. This type of farming is in decline due to its harmful impact on the environment.

Commercial plantations: this type of agriculture is labor- and capital-intensive. The major products cultivated in this way are tea, coffee, rubber, bananas, and palm oil. Plantations were originally developed to provide these inter-tropical crops for U.S. markets.

Green Revolution

The term "Green Revolution" was originally used in 1968 by former *United States Agency for International Development* (USAID) director William Gaud in 1968 to describe the spread of new technologies in the agricultural sector.

Today, the "*Green Revolution*" refers to research and development initiatives taking place from the 1930s through the late 1960s that increased worldwide agricultural production. These initiatives include:

- New chemical fertilizers to increase crop yields;
- Synthetic herbicides and pesticides to control weed growth and prevent destruction by insects;
- High-yielding varieties of cereal grains and other hybridized seeds;
- Expansion of irrigation infrastructure; and
- Modernization of management techniques.

Developing nations were the ones who benefited the most from the "Green Revolution." The "Father of the Green Revolution," Norman Borlaug (Nobel Peace Prize in 1970), was credited with saving over one billion people from starvation.

Genetic engineering and crop production

The ability to genetically modify and engineer crops has solved many agricultural problems. The composition of an organism is altered by inserting, deleting, or changing specific pieces of DNA. Genetic engineering can increase the productivity of crops by modifying them to prevent diseases; however, some of the diseases or insects that prey on crops can adapt to the genetic modifications designed to keep them away.

Genetically modified organisms (GMO) are causing concerns regarding their safety for human health and broader ecosystems, and many consumer groups are calling for more independent research and stricter regulations for production and labeling.

Deforestation

Deforestation is the process of clearing an area of trees with no intention of replacing them. A lack of tree and plant roots interferes with soil cohesion and a landscape's ability to capture, retain and transpire water; this causes most of the precipitation to accumulate on the soil's surface, increasing flash floods and landslides.

Destruction of plants results in the degradation of soil. As the soil loses its ability of infiltration, it becomes unfit for the growth of plants, making the land sterile.

Figure 4.2. *Deforestation of peat forest in Indragiri Hulu, Sumatra, Indonesia*

Many organizations have stated that the major reason for deforestation is subsistence farming and logging. Biofuel development, such as growing sugarcane for ethanol and Jatropha plant for biodiesel, has also led to deforestation.

As a result of deforestation, Central America has lost two-thirds of its tropical lowland forests, and South America has lost 70% of its rainforests. Brazil has lost a large portion of its rangelands, and, in certain areas of Brazil, cattle farming has brought about extensive destruction of rangelands. As a result, Brazil has declared deforestation a national emergency

Irrigation

Irrigation is how water is supplied to crops for them to thrive in areas where there is no water source or where water is inconsistent. Irrigation is practiced in many ways, such as through surface irrigation, localized irrigation, drip irrigation, sprinkler irrigation, and rotary irrigation. The innovation of irrigation has allowed for increased agriculture and crop growth, and it has been essential to the development of human society throughout history.

Sustainable agriculture

Sustainable agriculture is a type of farming that focuses on crop production with minimal effects on the environment. Sustainable agricultural practices utilize 30% less energy than industrialized agriculture.

Controlling pests

Pesticides are chemical agents used to control pests by killing them or inhibiting their growth ("– *icide*" means "*to kill*"). Pests have always been a problem for farmers; the quality and yield of crops can be enhanced by limiting the infestation of pests and the diseases they spread as a result.

The International Plant Protection Convention and Phytosanitary Measures Worldwide (2012) defines a *pest* as "*any species, strain or biotype of plant, animal or pathogenic agent that is injurious to plant or plant products.*"

Pest control is important for the regulation and management of unwanted species, which can cause detrimental impacts on livestock, disturb ecology, or create an adverse effect on the economy.

Types of pesticides

The main types of pests are weeds (unwanted plant species that compete with crops for nutrient supply), insects, rodents, and microorganisms (such as bacteria and fungi). Pesticides are categorized as chemical pesticides (which come from synthetically derived chemicals) or biopesticides (which come from naturally-occurring chemicals originating from plants or animals).

Different pesticides are used to target specific groups of pests:

- Insecticides: agents used to kill insects (e.g., DDT, gammexane, etc.)

- Miticides: agents used to kill ticks and mites (e.g., Permethrin, organophosphates, etc.)

- Fungicides: agents used to kill and inhibit the growth of fungi that cause molds, rots, and diseases in plants (e.g., Thiram, Bordeaux mixture, etc.)

- Rodenticides: agents used to kill rats, mice, moles, and gophers that damage crops (e.g., Aluminum phosphide, etc.)

- Pheromones: biologically active chemicals used to attract insects or disrupt their mating behavior. These chemicals mimic the natural molecules produced by insects and are used to draw them into traps.

- Herbicide/weedicide: agents used to prevent the growth of weeds (e.g., Benzipram, benzadox, etc.)

Costs and benefits of pesticide use

Pesticide production is a $32 billion industry; more than 5 billion pounds of pesticides a year are applied to crops worldwide, with 1 billion pounds being applied in the U.S. alone. Benefits include increased crop yield per acre and greater profits for farmers (both in having more products to sell and in saving money on labor costs).

However, these relatively short-term benefits have long-term consequences to human and environmental health, which indirectly costs the U.S. at least $8 billion a year.

Figure 4.3. *Pesticide application, Yuma, Arizona*

Many pesticides are proven neurotoxins, as well as likely carcinogens and endocrine disruptors. Strong evidence links pesticide exposure to congenital disabilities and impaired fertility in humans. Worldwide, there are an estimated 1 million human poisonings a year via pesticides, with approximately 20,000 deaths.

In the U.S., there are about 67,000 poisonings reported, most of them non-fatal; estimates indicate that this is considerably less than the actual total. These poisoning totals do not consider the long-term, low dose exposure many humans receive by consuming produce treated with pesticides. Thousands of domestic animals are poisoned by pesticides each year; meat, milk, and eggs are also contaminated.

Pesticides can lead to the destruction of natural predators and parasites and create secondary outbreaks of those pests whose natural enemies were destroyed by the initial round of chemicals (leading to additional pesticide applications). Pesticide use can also create pesticide-resistant pests. Those pests naturally adapted to a control agent reproduce preferentially, changing the genetic composition of the pest population.

Genetic variation means some pests within each species are naturally less vulnerable to specific control agents. They survive attempts at control, while the vulnerable population is killed off. Survivors reproduce in greater numbers and change the genetic makeup of the pest population by eliminating the susceptible genes from the population.

Pesticides can contaminate groundwater and well water, where chemical residue can remain for long periods. By seeping into aquatic ecosystems, pesticides can kill off fish, fish eggs, and fish food; they can also contaminate living fish to the point of becoming unfit for consumption. Birds are also damaged by pesticide exposure.

DDT (dichlorodiphenyltrichloroethane) nearly wiped out the bald eagle in the mid-20th century. While not lethal to adult eagles, DDT interfered with their fertility and calcium metabolism, producing eggs with thin, brittle eggshells unable to withstand their parent's weight.

Much of the agricultural industry depends on bees for the pollination of fruit and vegetable crops. Most insecticides are toxic to bees and have a major negative impact on their populations. It is estimated that 20% of all losses of honeybee colonies are from pesticide exposure; some pesticides can kill adult bees outright, while others make them severely weakened or affect the development of young bees.

New research suggests that two pesticides (neonicotinoids and coumaphos) target a bee's brain and nervous system, interfering with its ability to learn and remember.

Integrated pest management

Integrated pest management (IPM) is a program that aims to minimize pest infestation and long-term pest control by utilizing techniques that cause a minimal level of harm to the existing ecosystem.

The key focus of such a process lies in identifying and controlling the environmental factors and conditions in which pest infestation takes place.

Factors such as humidity, temperature, and waterlogging play an important role in pest management. If these factors are closely monitored, it is possible to control pest infestations with greater success, thereby preventing or reducing crop damage.

The University of California has introduced a course in IPM and issues statewide guidelines for minimizing the problems associated with pest infestations.

The five major components of an IPM program are:

1. Pest identification;

2. Monitoring and assessing pest numbers and damage;

3. Guidelines for when management action is needed;

4. Preventing pest problems; and

5. Using a combination of biological, cultural, physical/mechanical, and chemical management tools.

Relevant laws

The United States government works together with the *Center for Disease Control* (CDC) to implement effective IPM practices and ensure the safety of both consumers and agricultural workers.

The CDC helps protect public well-being by providing information on crop health and potential disease outbreaks within animal populations.

The Environmental Protection Agency (EPA) and various pest management institutions provide public information and tools for handling rodents, mites, and other pests, and they closely monitor the deleterious effect these organisms have on humans and the food they consume.

Many pesticides have been banned in the U.S. due to health concerns. The most famous harmful pesticide is DDT, which was widely used after World War II both agriculturally and to eradicate malaria.

Figure 4.4. *DDT solution sprayed onto sheep for tick control in Benton County, Oregon, 1948*

Concerns over DDT's adverse effects on humans, wildlife and the environment came to a head in the 1960s, spurred by the publication of Rachel Carson's *Silent Spring*. After countless committees, lawsuits, hearings, and appeals, DDT was banned in 1972.

However, the chemical has long-lasting effects, and DDT breakdown products can be found in soil, crops, and human bloodstreams even to this day.

In 1985, the highly toxic insecticide *aldicarb* was responsible for over 2,000 cases of poisoning in the U.S. (including six deaths and two stillbirths) due to contaminated melon. However, it took until 2010 for legislation to be passed that would phase out aldicarb's production by 2015.

Notes

Forestry

The United States has approximately 751 million acres of forest, which is about one-third of the country's total land area. National, state, and local governments own about 43% of all U.S. forests; the rest are owned by private landowners, including more than 22 million family forest owners. The U.S. has about the same amount of forested land today as it did 100 years ago, despite a nearly three-fold increase in population over the same period (National Report on Sustainable Forests – 2010). This is a striking contrast to many countries, where wide-scale deforestation remains a pressing concern. The United States and Canada together contain about 15.5% of the world's total forests.

Forestry as a profession in North America is about 100 years old. Over the past century, the field has evolved from practices focused on maximizing timber value to approaches deeply rooted in ecology, science, and sustainability. Modern-day forestry practitioners complete rigorous college programs and participate in continuing education, certification, and licensing programs to establish and maintain their professional credentials. These programs are as demanding as those undertaken by aspiring engineers or architects. Laws addressing safety and workers' rights also govern forestry activities.

Figure 4.5. *A forest on San Juan Island, Washington.*

About 25% of forests in the U.S. are designed for the protection of soil and water and the conservation of biodiversity, including more than 100 million acres of reserves and roadless areas. The remaining 45% is designated for multiple uses, and it is often referred to as *working forest*. These lands are cared for by public and private interests that balance their use as a source of income with the objectives of protecting wildlife, maintaining water quality, providing recreation, and maintaining the forests' aesthetic value.

In 2003, the United States performed a comprehensive assessment of the forest practice regulatory programs within state governments. An extensive review of the literature was undertaken, and information was collected from program administrators in all 50 states. The assessment determined that a wide range of forestry practices be implemented in private forests.

Many state agencies control forestry practices and are substantially invested in related programs, with especially stringent regulatory programs in 15 states. Administrators have suggested that regulatory program design would benefit from research focused on identifying forestry sectors that require attention, determining the best means for equitably enforcing regulation and evaluating program performance, and designing information management systems to use for monitoring regulatory programs.

National forests

The U.S. Forest Service oversees the protection of 155 National Forests, which contain a total area of almost 190 million acres — 8.5% of the total land of the United States. There are two distinctly different types of National Forests. Those east of the Great Plains are primarily re-acquired or replanted forests, meaning that the land was purchased from private owners by the United States government to create new National Forests.

The National Forests west of the Great Plains have always been held by the U.S. government (or were donated to it). The Yellowstone Park Timber and Land Reserve were designated as the first National Forest on March 30, 1891, following the *Forest Reserve Act* of 1891, which allowed the president to create forest reserves for the enjoyment of the public.

The 1897 *Organic Act* outlined more specific protection purposes for which forest reserves could be established, such as securing water supplies and supplying timber. Natural resources are allowed and sometimes encouraged, to be extracted from National Forests. Forest reserves became part of the U.S. Department of Agriculture's recently established *Forest Service* under the *Transfer Act* of 1905. As of September 30, 2014, the Forest Service oversees 193,062,995 acres of land, most of which have been designated as National Forest.

Forty states have at least one National Forest. Alaska has the most National Forest land (21.9 million acres), followed by California (20.8 million acres) and Idaho (20.4 million acres).

Timber companies and environmentalists often disagree about how National Forest land should be used. The disagreements center on endangered species protection, logging of old-growth forests, intensive logging, undervalued stumpage fees, mining laws, and roadbuilding.

Forest management

Forest management is a branch of forestry concerned with the overall economic, legal, scientific, and social aspects of forests. This includes management for aesthetics, recreation, water, wildlife, wood products, and genetic resources.

Management can be based on conservation, economics, or a mixture of the two. Techniques include timber extraction, planting of various floras, cutting roads and pathways through forests, and preventing fire.

Forest management in the United States operates under layers of federal, state, and local regulations and guidelines that foresters and harvesting professionals must follow to protect water quality, wildlife habitat, soil, and other resources.

For decades, net forest growth has outpaced the amount of wood harvested; this supports the idea that landowners who are economically dependent on wood have a strong incentive to continue sustainable management practices. This aligns with global forest data, which indicates that forest products and industrial round wood demands provide revenue and policy incentives to support sustainable forest management.

However, due to the expansion of urban development and increased demands from other interests vying for more land, it is important to ensure that landowners will continue to have reasons to prevent deforestation.

About 30% of the forest area of the United States is classified as *production forest*, where the land is managed primarily to harvest forest products. Methods that are currently employed to manage and harvest trees include:

- *Even age management*: essentially, the practice of tree plantations;

- *Uneven-age management*: maintain a stand with trees of all ages from seedling to mature;

- *Selective cutting*: specific trees in an area are chosen and cut;

- *High grading*: cutting and removing only the largest and best trees;

- *Shelterwood cutting*: most trees are removed except for scattered, seed-producing trees used to regenerate a new stand;

- *Clear-cutting*: all the trees in an area are cut at the same time (this technique is sometimes used to cultivate shade-intolerant tree species); and

- *Strip cutting*: clear-cutting a strip of trees that follows the land contour; the corridor is allowed to regenerate.

Tree plantations

There are two categories of planted trees recognized by the Forest Stewardship Council's U.S. Forest Management Standard: conventional plantations and "Principle 10 Plantations." Conventional plantations, which account for most planting projects in the U.S., are treated the same way as natural and semi-natural forests under this standard.

Special considerations apply to Principle 10 Plantations, which are strictly controlled and made up of blocks of trees that lack the traits of natural forests; this includes exotic trees (such as Eucalyptus) and cloned trees lacking genetic variation.

The FSC-US Standard aims to prevent further loss of natural and semi-natural forests to Principle 10 Plantations. While Principle 10 Plantations can take some of the pressure off natural forests in producing commercial products, they generally lack biodiversity and are not as good for harboring wildlife.

Management practices that could effectively disrupt the main traits of native forest and result in a Principle 10 Plantation tag include:

- Harvest cycles short enough to prevent stands from developing natural understory stages;

- Steady use of chemical herbicides and/or frequent fertilization;

- Practices that promote a single species on sites normally occupied by multiple-species forests; and

- Failing to leave at least a minimal number of trees or undisturbed spots for the benefit of wildlife.

The *American Tree Farm System* (ATFS) is the largest and oldest woodland certification system in America. It is one of three certification systems in the United States (the others include the FSC and the Sustainable Forestry Initiative).

ATFS specializes in private forests, primarily those held by individuals and families; it currently certifies over 24 million acres of forestland. This network of over 90,000 woodland owners is organized through state committees and governed at the national level. 45 of the 50 states have committees; Alaska, Arizona, Hawaii, North Dakota, and Utah currently do not have programs. With national coordination, ATFS strives to "*work on-the-ground with families . . . to promote stewardship and protect the nation's forest heritage.*"

According to the Standards of Certification for ATFS, woodland owners must own 10 or more acres and have a plan for management that involves recognizing wildlife habitats, protecting water quality and threatened/endangered species, and practicing sustainable harvest. The minimum acreage to qualify for a tree farm refers to woodland.

So, acreage that includes grazing or other non-wooded lands must have at least 10 acres of forest to qualify. Furthermore, programs in different areas that support tree farming activities may require larger forested acreages as well as additional criteria.

For example, The Forest Ag Program in Colorado requires the following standards:

- The landowner must perform forest management activities to produce tangible wood products (e.g., Christmas trees, firewood, fence posts) for the primary purpose of obtaining a monetary profit;

- The landowner must have at least 40 forested acres; and

- The landowner must submit a Colorado State Forest Service-approved management plan prepared by a professional forester or natural resources professional.

As a program of the *American Forest Foundation* (AFF), the American Tree Farm System focuses on the long-term sustainability of America's forests in ecological and economic terms.

AFF's vision statement is: *"AFF is committed to creating a future where North American forests are sustained by a public that understands and values the social, economic, and environmental benefits they provide to our communities, our nation, and the world."*

Old-growth forests

Old-growth forests largely contain trees over 30 inches in diameter that are often hundreds, or even thousands, of years old and possess complex canopies. Much of the world's old-growth forests have been cleared; according to the *World Resources Institute* (WRI), only 21% of the original old-growth forests remain.

Differences in the methods used to inventory remaining stands of *old-growth* forest can produce major discrepancies. In 1991, for example, the U.S. Forest Service and the nonprofit Wilderness Society each released their inventory of old-growth forests in the Pacific Northwest and Northern California. They both used the Forest Service's definition of *old-growth* based on the number, age, and density of large trees per acre, the characteristics of the forest canopy, etc.

However, because each agency used different remote sensing techniques to collect data, the Forest Service came up with 4.3 million acres of old-growth, while the Wilderness Society found only 2 million acres. The *National Commission on Science for Sustainable Forestry* (NCSSF) also studied the data and concluded that 3.5 million acres (or 6%) of the region's 56.8 million acres of forest qualified as old-growth.

Figure 4.6. *A yellow birch in the Allegheny National Forest, Pennsylvania.*

In the Northeast, less than 1% of forests are old-growth, though mature forests that will become old growth in a few decades are more abundant there than elsewhere in the country.

The Southeast has even less acreage of old-growth; only 0.5% of the total forest area is considered old-growth, distributed between 425 sites across the region. The Southwest has only a few scattered pockets of old-growth, mostly Ponderosa pine, and is not known for its older trees. Old-growth forests are even scarcer in the Great Lakes region.

The protection of old-growth trees is vitally important. These forests provide critical habitat for salmon since they perform some functions that help foster salmon populations. Trees and vegetation that line freshwater streams give the shade needed for spawning streams.

Root systems stabilize stream banks and help slow erosion. Leaves and needles that fall into the water provide another food source (as they often carry small invertebrates) and help maintain water pH. Trees and branches that fall into streams provide shelter for young fish and offer an additional food source since the organic matter they contain sustains aquatic insect populations. Furthermore, they can direct and shape advantageous streamflow.

Also, the loss of forested areas due to unsustainable human practices exacerbates the current climate change crisis. Greenhouse gases released by human activity, which would normally be mitigated by the ability of trees to absorb CO_2 emissions, increasingly escape into the atmosphere as deforestation continues.

Between 2000 and 2005, deforestation in tropical regions accounted for roughly 12% increase in greenhouse gas emissions.

Forest fires

A wildfire is an uncontrollable fire that occurs in an area where there are combustible vegetation and the presence of outside ignition factors (e.g., humans, lack of moisture). A wildfire differs from other fires by its extensive size, the speed at which it can spread from its source, its potential to unexpectedly change direction, and its ability to jump roads and rivers.

Figure 4.7. *The Brins Fire, Sedona, Arizona, 2006.*

Four of the worst fires in U.S. history all broke out in the Upper Midwest during the week of October 8-14, 1871. The most prominent of these was the Great Chicago Fire, which left over 100,000 people homeless and caused massive economic disruption.

Simultaneously, the Great Michigan Fire ravaged the cities of Holland and Manistee, while across the state another fire destroyed the city of Port Huron. The Great Peshtigo Fire in Wisconsin may have been the most devastating of these fires. It killed more than 1,500 people — the most fatalities of any fire in U.S. history. It is unclear whether these concurrent fires were simply coincidence, or whether they were the result of a natural phenomenon, such as a meteor shower or high winds moving through the region.

Some of the worst forest fires in recent years are:

- May 9-July 23, 2012, in Catron County, New Mexico. 297,845 acres of the Gila National Forest was burned;

- June 30, 2013, in Prescott, Arizona. The Yarnell Hill Fire killed 19 firefighters (the greatest single loss of firefighters since 9/11) and burned 8,000 acres of land;

- July 14-August 7, 2014, in Washington. The largest wildfire in the state's history, started by lightning, burned 250,000 acres of land; and

- September 2015 in California. The Valley Fire burned 73,000 acres of land, the Butte Fire 70,000 acres, and the Rough Fire 14,000 acres.

While wildfires can be devastating to property and human life, fire is an important part of many ecosystems, including prairie, savanna, and coniferous forest. Many plant species in these environments (such as Giant sequoias) require fire to germinate and thrive.

By suppressing naturally occurring fires in these habitats, humans threaten both these plant species and the animals that depend on them. Suppression also results in a build-up of flammable debris, which can result in wildfires that are even more destructive than normal.

Prescribed burning, a technique that utilizes smaller, controlled fires in the cooler months, can help to prevent these issues. Prescribed burns can reduce fuels, improve wildlife habitat, control competing vegetation, improve short-term forage for grazing, help control tree disease, and perpetuate fire-dependent species.

Rangelands

Rangelands are extensive areas of terrain with few trees, where the native vegetation is predominantly grasses, grass-like plants, forbs, or shrubs. Rangeland environments include prairies, grassland, savanna, and steppe. Rangelands are generally unsuited for crop cultivation but are good for grazing livestock. The primary difference between rangeland and pasture is management; rangelands tend to have natural vegetation (along with a few introduced plant species) only managed by grazing, while pastures have forage adapted for livestock and managed by seeding, mowing, fertilization and irrigation.

Figure 4.8. *Rangeland of the Red Desert, Wyoming.*

The U.S. has about 770 million acres of rangelands; over half of this total is under private ownership. Nearly 43% of these rangelands belong to the federal government, and the rest are held by local and state governments (National Research Council, 1994).

Rangelands are not just good for grazing but produce a wide variety of goods and services, including wildlife habitat, water, mineral resources, wood products, and recreation. Though the U.S. government has tried to preserve these rangelands, their destruction still takes place for several reasons.

Overgrazing

Overgrazing has contributed to the destruction of 48% of rangelands. Overgrazing arises when grasses are continuously eaten by livestock without periods of recovery. Rangelands predominantly consist of perennial grasses, which can be renewable when not overexploited. Normal grazing allows the grass time to recover and regrow via the metabolic reserves in what remains of its blades and stems.

When grass is grazed too severely, and little to no stem remains, the grass uses energy stored in its roots to support regrowth. This causes root dieback, which in turn causes soil erosion and compaction, making it even harder for regrowth. Overgrazing decreases the rate of growth of desirable plants, often up to 90%. In these harsh conditions, undesirable weeds thrive.

New Mexico's South Chiflo Management Area has become bare due to a history of overgrazing. The area once flourished with lush green grasses and many small trees. More than 30,000 sheep were allowed to graze in this area in the 1930s. Today, this rangeland still has numerous bare patches as a result of that overgrazing.

The use of sustainable grazing techniques will enrich the soil, increase the growth of desirable plants, decrease the invasion of weeds, and result in thriving rangeland. However, due to the private ownership of many rangelands, it is difficult for rangeland management teams to take the necessary steps to reduce overgrazing.

Ranchers also over-exploit public rangelands as grazing grounds for livestock. Cattle-feeding has gained importance in the United States of America, and cattle are fed vigorously on and off the rangelands as ranchers seek increased income.

Deforestation

Deforestation is the conversion of forested areas to non-forested areas. Natural deforestation can be caused by tsunamis, forest fires, volcanic eruptions, and glaciations. Humanmade deforestation in rangelands is carried out for two primary reasons: to create pastures and to convert rangeland to farmland. These non-forested areas are used as grasslands for grazing livestock, grain fields, mining, petroleum extraction, commercial logging, and urban sprawl.

Deforestation results in a degraded environment with reduced biodiversity and reduced ecological services. Deforestation threatens the extinction of species with specialized niches, reduces the available habits for migratory species of birds and butterflies, decreases soil fertility, and allows runoff into aquatic ecosystems. It also causes changes in local climate patterns and increases the amount of carbon dioxide released into the air from burning and tree decay. In addition to the direct effects brought about by deforestation, indirect consequences caused by edge effect and habitat fragmentation can also occur.

Deforestation alters the hydrologic cycle, potentially increasing or decreasing the amount of water in the soil and groundwater. This, in turn, affects the recharge of aquifers and the moisture in the atmosphere.

Shrinking forest cover lessens the landscape's capacity to intercept, retain, and transport precipitation. Instead of trapping precipitation, which then percolates to groundwater systems, deforested areas become sources of surface water runoff, which moves much faster than subsurface flows. The faster transport of surface water can lead to flash flooding and more extensive flooding than would occur with the forest cover.

Deforestation also contributes to decreased evaporation. This lessens atmospheric moisture and precipitation levels. It also affects precipitation levels downwind from the deforested area as the water is not recycled to downwind forest but, instead, is lost in runoff and returns directly to the oceans.

Forests are also important carbon sinks. Forests can extract carbon dioxide and pollutants from the air, contributing to biosphere stability and reducing the greenhouse effect. Forests are also valued for their aesthetic beauty and as a cultural resource and tourist attraction.

There are only 10,877 (thousands of sq. km) forests present in North America, 1,779 (thousands per sq. km) present in Central America, and 9,376 present in South America. Deforestation has depleted nearly 60% of the rangeland cover.

The Atlantic coast of Brazil has lost nearly 90-95% of its forest cover. There is no strong management of the rangeland resources in the United States, and the governance structure is not up to standard. Illegal deforestation is also prevalent in many parts of the United States.

Desertification

Desertification occurs when relatively dry (but productive) land becomes increasingly arid, which generally results in loss of plant life, animal life, and bodies of water. Dryland ecosystems cover nearly half of Earth's land area, and because of the natural scarcity of water, they are especially vulnerable to desertification; 10 to 20% of drylands are already affected.

Dryland covers most of the western half of the continental U.S.; one-third of this is already severely desertified, particularly in Arizona and on the Navajo reservation in New Mexico. In this region of the United States, the main cause of desertification has been overgrazing and water erosion.

The increasing salinity of the Colorado River may lead to the eventual desertification of the surrounding area. The rangelands of Mexico and about 22% of South America have been severely desertified. UNESCO began efforts to prevent desertification in 1962.

Figure 4.9. *Dried out soil, Sonora, Mexico*

Rangeland management

The goal of rangeland management is to implement improvement plans to conserve and protect rangelands. Governmental agencies have determined that brush control programs help protect rangelands and benefit the ecosystem.

Rangelands are one of the main ecosystems responsible for storing carbon. Their destruction releases that carbon back into the atmosphere; increased atmospheric carbon dioxide is a leading cause of climate change.

Management teams are increasing the density of shrubs in and around rangelands, as their root systems help water retention and prevent erosion and may be crucial to rangeland restoration. Officials have asked ranchers to reduce activities that cause soil erosion and decrease the fertility of the soil on rangelands.

The *United Nations Convention to Combat Desertification* (UNCCD) was established in 1994 to reverse the effects of desertification in rangelands.

Federal rangelands

The federal government manages about 33 million acres of rangeland. These federal rangelands are managed so well that they provide pure water, produce renewable energy sources, and have become centers of recreation for the public.

In the western United States, livestock can graze openly on federal land if the rancher pays fees. There is a set limit on the amount of grazing that can take place on federal rangelands; those who exceed it (or who do not pay the fee) are subject to a fine and/or imprisonment.

The U.S. government aims to protect the 39% of federal rangelands, which are on the verge of desertification. Through the *Federal Land Policy and Management Act* of 1976, the government puts 50% of the grazing fees it collects toward the rehabilitation, protection, and improvement of rangeland.

The state of Arizona has been declared a protected area due to its lack of current rangeland, in the hope that this ecosystem can be restored in some areas of the state.

Notes

Other Land Use

The term *land use*, as defined by Albert Z. Guttenberg in his article, "A Multiple Land Use Classification System," published in the *Journal of the American Planning Association*, 25:3, 143-150 (1959) reads:

"Land use is characterized by the arrangements, activities, and inputs people undertake in a certain land cover type to produce, change or maintain it."

Land use also includes near-surface water.

Any given land area is often used to meet multiple objectives. Land use involves the modification or conversion of natural environments into human-built environments.

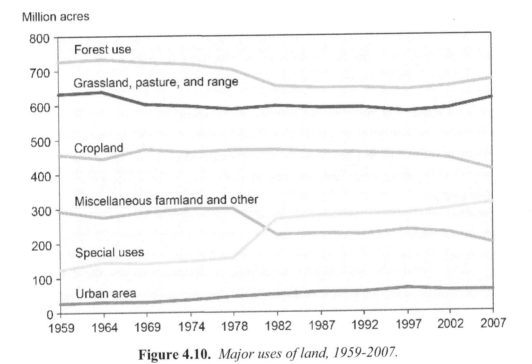

Figure 4.10. *Major uses of land, 1959-2007.*

Urban land development

The increase in the proportion of people living in urban areas is a global phenomenon. Many cities struggle to cope with the challenges of urbanization; these include increased poverty, slum growth, climate change, and resource scarcity.

Urban land use planning is based on sustainable development principles, which require well-managed, well-planned initiatives.

Urban land development deals with various issues, such as reducing the negative impacts of human-built environments on climate and natural resources, protecting built environments from unpredictable and volatile conditions, sensible use of energy resources, and making urban real estate more attractive, investment-wise.

Planned development

The planned development of urban areas shows the incredible changes stemming from new ideas and new technology, which can produce sustainable mega-infrastructures. Urban planning guides ensure the orderly development of new built-in and satellite communication, which can communicate in and out of urban areas.

Urban planning encompasses architecture, landscape, civil structures, and public administration so that strategic, policy, and sustainability goals can be achieved. The development of urban areas has become professionalized over the years. The Town and Country Planning Association was founded in 1899; it has grown along with the growth of urban centers.

Suburban sprawl

According to the UCL Center for Advanced Spatial Analysis, the definition of *suburban sprawl* (sometimes referred to as *urban sprawl*) is *"uncoordinated growth; an indication of the expansion of community without concern for its consequences."*

It is the expansion of human populations away from central urban areas into low-density, usually car-dependent bedroom communities, where commuters often make up a large portion of the population.

Often, this leads to single-use development, a situation where commercial, residential, and industrial areas are separated from one another to the extent that walking, public transit, and bicycles become impractical, requiring residents to own cars.

Figure 4.11. *Suburban sprawl, Rio Rancho, New Mexico.*

Urbanization

Urbanization is the gradual migration of populations from rural areas to urban ones throughout history. It has several components, including geography, sociology, urban planning, economics, and public health. Urbanization has occurred at different rates in different countries. For example, the U.S. and the United Kingdom are more urbanized than India, Swaziland, and Niger.

Transportation infrastructure

Transportation is defined as the movement of people, animals, and goods from one location to another using air, road, rail, or cable. *Transportation infrastructure* is a fixed installation that provides logistical support for transport. Examples include roads, railways, airways, waterways, canals, pipelines, and terminals such as airports, railway stations, and seaports. Transport is essential to modern society; however, contemporary means of transportation require a large amount of energy consumption. Finding sustainable means of transportation is essential to humans' continued progress as a species.

Federal highway system

The *Federal-Aid Highway Act* of 1956, signed into law by President Eisenhower, created a 41,000 mile "national system of interstate and defense highways," which would provide safe and quick transcontinental travel and eliminate unsafe roads, traffic jams, and inefficient routes. Under this system, the federal government paid 90% of the cost of the expressway construction. It was declared that expressways should be at least four lanes wide and designed for high-speed driving.

Canals and channels

Canals and channels are human-made waterways that control the flow of water. The main difference between them is that canals cut across drainage divides, while channels run parallel to rivers and share their drainage basins. There are two main types of canals, those used as waterways and those used to facilitate the water supply. Canals constructed for use as waterways are often connected to lakes and rivers and flow from one populated area to another to enable quicker travel.

Water supply canals are used for the conveyance and delivery of water for human use, industrial use, hydropower plants, or agricultural irrigation. Due to globalization, canals are becoming increasingly important; this has led to the digging of new canals and the expansion of existing ones. The two best examples of canals in the modern era are the Panama Canal, which connects the Pacific Ocean to the Atlantic Ocean in Panama, and the Suez Canal, which connects the Red Sea to the Mediterranean through Egypt.

Figure 4.12. *USS America aircraft carrier passes through the Suez Canal, Egypt, 1981.*

Roadless areas

Access roads into wilderness habitats, while providing convenient entry for industry and recreational activities (e.g., fishing, hunting, sightseeing), can lead to a host of negative effects such as erosion, pollution, and loss of biodiversity. Also, the building of roads can lead to further development of "splinter roads," which can cause even further fragmentation and environmental decay in ecologically delicate areas.

In 2001, the U.S. Forest Service, after nearly three years of study and analysis, passed the *Forest Service Roadless Area Conservation Rule* (Roadless Rule), which preserves 58.5 million acres of pristine National Forests and Grasslands (about one-third of all NFS lands) from most logging and road construction. It advocated that roads should be constructed to avoid harm to the surrounding wilderness and wildlife. These areas provide critical habitat for more than 1,600 threatened, endangered, or sensitive plant and animal species.

Ecosystem impacts

Historically, the impact of land use on ecosystems has been underestimated and ignored. However, in recent years, it has become a growing concern. Some land-use practices can degrade the quality of soils, waterways, air, and other natural resources.

One land-use issue that often arises is the "carbon footprint" created by increased food and bioenergy production. Ecosystem impact assessments are used by authorities, industries, and individuals to determine how the environment has been affected by land use.

Public and federal lands

In all modern nations and states, the land which is held or owned by the central or local government is called *public land.* Federal lands are those in the United States that are owned by the U.S. federal government. According to a survey conducted by the Department of the Interior, out of the 2.27 billion acres of land in the U.S., around 28% is owned by the government.

Management

The management policy for public and federal lands was established by the *Federal Land Policy and Management Act* of 1976. This established the concept of *multiple-use* land, which is defined in the Act as *"the management of public lands and their various valuable resources so that they are used in such a manner that will best meet the present and future needs of the American people."*

Wilderness areas

The *National Wilderness Preservation System* of the U.S. protects and manages wilderness areas to preserve their natural and environmental conditions. *Wilderness* is defined as an area where nature is allowed to exist unimpeded by human interruption. There are 762 acres designated wilderness areas, which total nearly 109 million acres of land. Wilderness areas include wildlife sanctuaries, wildlife refuges, national parks, and national forests. As urbanization leads to the expansion of cities, it is important to consider the impact this growth has on wilderness areas and the animals and natural resources they contain.

National parks

The United States has 59 protected areas known as *national parks*, which are operated by the *National Park Service* (NPS). The NPS was created *"to conserve the scenery and the natural and historic objects and wildlife therein and to provide for the enjoyment of the same in such manner and by such means as will leave them unimpaired for the enjoyment of future generations."*

The first national park, Yellowstone, was signed into law by President Ulysses S. Grant in 1872. Criteria for the selection of national parks include natural beauty, unique geological features, unusual ecosystems, and recreational opportunities. The total area protected by national parks is approximately 51.9 million acres.

Figure 4.13. *Mount Saint Elias, Wrangell-St. Elias National Park and Preserve, Alaska.*

Twenty-seven states have national parks, as do the territories of American Samoa and the United States Virgin Islands. California has the most (nine), followed by Alaska (eight), Utah (five) and Colorado (four). The four largest national parks are all located in Alaska.

The biggest national park is Wrangell–St. Elias; at over eight million acres, it is larger than each of the nine smallest states. The smallest national park, at less than 6,000 acres, is Hot Springs in Arkansas. The most-visited national park is the Great Smoky Mountains in North Carolina and Tennessee, which had over 10 million visitors in 2014. In contrast, only 12,669 people visited the remote Gates of the Arctic in Alaska that year.

Wildlife refuges

Wildlife refuges are official territories created by the government or held by private owners that are designed to protect endangered animals. Hunting, predation, and competition are restricted in protected areas. The first government-owned refuge in the Americas was the Lake Merritt Wildlife Refuge in Oakland, California, which was established by then-mayor Samuel Merritt in 1870. Today, some national and international organizations have taken responsibility for protecting the existing system of nonprofit wildlife refuges. The American Sanctuary Association plays an especially vital role in protecting wildlife animals by providing accreditation to facilities that follow high standards to ensure that the animals in their charge are well-cared-for.

Forests

A forest is a large area of land covered with trees or other woody vegetation. Forests are the dominant terrestrial ecosystem of Earth; they account for 75% of the gross primary productivity of Earth's biosphere and contain 80% of Earth's plant biomass. In 2006, forests covered an area of 15 million square miles or approximately 30% of the world's land area.

Forests at different latitudes form distinctly different ecozones: boreal forests near the poles tend to consist of conifers, while tropical forests near the equator tend to be distinct from the temperate forests at mid-latitude. Their elevation and the amount of precipitation they receive also affect forest composition.

Wetlands

Wetlands are land areas that are saturated with water, either permanently or seasonally; this creates a distinct ecosystem of aquatic plants, such as mangroves and eelgrass. The main types of wetlands are swamps, which are forested, and marshes, which are dominated by herbaceous, non-woody plant species. Environment degradation is more prevalent in wetlands than in any other ecosystem on Earth. The main difference between wetlands and other classifications of land is that wetlands include aquatic plants that cannot be found elsewhere. The largest wetland located near an urban area is in Esteban Echeverria Partido, Argentina, and is called Laguna de Rocha.

Land conservation options

Land conservation is when steps are taken to protect natural landscapes and to return developed, disturbed area back to its natural state. Today, many people place much value on the legacy values and natural beauty of their land and often see it as important as financial concerns. Understanding conservation options is an important step in deciding the future of a land's use. Land conservation often results in positive financial outcomes through income or tax savings.

Preservation

Preservation is a process by which land and the natural resources it contains are deemed protected from human intervention to maintain the land's pristine form. Preservation does allow humans to access the land to enjoy its natural beauty and inspiration. One of the most famous land preservationists was John Muir, a Scottish immigrant who lived from 1838 to 1914. He had a great deal of admiration for California's Yosemite Valley and wanted to protect the land so people could enjoy it without stripping it of its natural resources. The influence of John Muir is still evident today through the continuing influence of his Sierra Club and the establishment of the Muir Woods National Monument, a preserved area of land in Northern California home to an ancient redwood forest.

Remediation

Remediation is the process of fixing a contaminated area using relatively nondestructive methods to clean and restore the area while causing a minimal amount of disturbance, damage and/or harm. Remediation utilizes various chemical, physical and biological methods of removing contaminants.

An interesting remediation method is known as bioremediation, where either naturally occurring or forcefully introduced organisms are used in an area to break down pollutants. This method was used to help clean up the Deepwater Horizon oil spill in the Gulf of Mexico in 2010. Naturally produced bacteria were used to help clean up; they were able to ingest and break down the oil into less harmful substances.

Figure 4.14. *Deepwater Horizon oil spill, Gulf of Mexico, 2010.*

Mitigation

Mitigation is the process of replacing a degraded site with a healthy site of equal ecological value in a different location. The purpose of mitigation is to compensate for destroying one area by purchasing or creating a new area of equal ecologic value. This method is less desirable than other land conservation methods because it still allows land to be destroyed, harmed, or damaged.

Restoration

Another technique used in land conservation is *restoration*, which is the process of returning ecosystems and their communities to their original natural condition. To restore an ecosystem, one must first know the condition the environment was in before it was altered.

Once the original condition is known, it is possible to attempt to return the area to its original state. This can be achieved by restoring waterways, reintroducing native plants and animals, and removing human-built infrastructure.

In the state of Florida, there is an ecosystem restoration project currently taking place in the Everglades, which has been rapidly drying up over the years due to irrigation and the development of flood management infrastructures. The project involves building a series of dams and other water control features to return the Everglades to its original state. If this is successful, it will be useful to the region because it might allow wildlife to return and increase ecotourism in the area.

Sustainable land-use strategies

Sustainable land use is an emerging and fast-growing design reform movement, which combines the creation and enhancement of land with the need to build high-performance infrastructures and buildings. Some components of sustainable land use implemented in cities and towns include open green space, walkability and connectivity, and sustainable water sources.

Mining

Mining is the extraction of intrinsically valuable minerals or geologically valuable materials from the Earth. Mining has been a part of the United States since the colonial era; however, there have been drastic changes in the type of minerals mined and the conditions of mining over the years.

The first extensive U.S. mining was carried out in North Carolina in 1799 and was called the "Carolina Gold Rush." Today, mining takes place in many states, including Arizona, Colorado, Minnesota, Alaska, California, Michigan, and Nevada. The U.S. plays a major role in the mining of coal, gold, silver, uranium, and copper.

Mineral formation

Minerals are naturally occurring, inorganic solids with definite chemical composition, and are found all over the world. The Earth contains a variety of unique minerals. Different minerals are used for many different purposes, including the construction of buildings, phone and computer components, jewelry, and fertilizers.

Some minerals are formed from saltwater when it evaporates from the Earth's surface, and some minerals are formed by the mixtures of water that seep through rocks far below the Earth's surface. Minerals are also formed when mixtures of molten rock cool down.

There are two main methods of mineral formation: formation from magma and lava and formation through chemical processes.

Formation from Magma and Lava

There are many places beneath the Earth's surface where temperatures are high enough to melt rock. The name for this melted rock is magma, which can reach temperatures above 1,000 °C. This magma does not always remain inside the Earth; sometimes it moves up to the surface through volcanoes, at which point it is known as lava.

When molten rock cools, it produces minerals like quartz and feldspar. Since magma cools at a much slower rate than lava, it produces minerals with larger crystals than those produced by lava.

Formation from solutions

Most of the water on Earth contains minerals. The minerals within the water are too small to be seen by the naked eye and are difficult to remove even when the water is filtered; however, there are other techniques by which these mineral deposits can be removed from the water.

Mineral formation from saltwater is a natural method of extracting minerals from a water solution. Mono Lake in California and Utah's Great Salt Lake contain enough saltwater to precipitate out minerals from the water through evaporation. When the water evaporates, it leaves behind solid particles of minerals that do not evaporate.

Some of the minerals remain in the water. If the amount of minerals in the water is too high, then the minerals come together to form mineral solids, which remain underwater. In Mono Lake, the water has too much calcium, so calcite deposits cause limestone "tufa" towers on the Earth's surface.

When magma flows, it heats the nearby water, which moves through cracks inside the Earth. Hot water can hold more solid particles than cold water, so when the hot water reacts with the rocks around it, it becomes saltier. This salty hot water then flows through cracks in the rocks called veins; the solid particles are deposited in these cracks in crystalline form. Examples of minerals that form in this way include dolomite, galena, fluorite, and gypsum.

Extraction of minerals

There are two basic types of extraction: underground mining and surface mining.

Underground Mining

There are two types of minerals located deep inside the Earth: a soft rock (e.g., coal) or a hard rock (e.g., copper, lead). Depending on the type of mineral one is attempting to mine, there are various methods by which to extract the underground minerals. Also, there are external factors (e.g., geological, economic, safety) that affect the mining methods that can be used.

U.S. Mining Fatalities CY 1978-2014

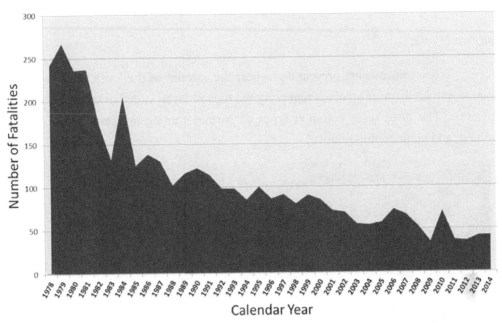

Figure 4.15. *U.S. Mining Fatalities CY 1978-2014. U.S. Department of Labor, Mine Safety, and Health Administration. 2016.*

In hard rock mining, blasting is used to separate the mineral from its deposits and to dislodge the waste rock. Ventilation is important in hard rock mining. Any toxic gases produced by blasting may become trapped in the mine and can prove harmful to miners or the structural integrity of the mine; many of these gases are highly combustible.

Also, it is important that the walls and the opening of the underground mine be structurally sound, so that sudden collapses that can trap, injure, or kill miners don't occur. After all useful, cost-effectively accessed minerals are removed from a mine, it is sealed off and either intentionally collapsed or left to collapse on its own.

In the soft mining method, longwall and room-and-pillar mining methods are most widely used. In both methods, automation and ventilation are provided for. Room-and-pillar is a system in which the mined material is extracted across a horizontal plane, creating arrays of rooms and pillars.

Pillars of untouched material are left to support the roof overburden, and open areas or "rooms" are extracted. In the longwall method, coal is mined in large, single slices. The equipment consists of a coal shearer mounted on a conveyor underneath a series of self-advancing hydraulic roof supports. Almost the entire process can be automated.

Surface Mining

When minerals are present on or near the exterior of the Earth, *surface mining* is used to extract them. It involves removing the topsoil layer, called overburden, to recover minerals. The three most common types of surface mining are open-pit mining, strip mining, and mountaintop removal.

Figure 4.16. *Open-pit mining, Sunrise Dam Gold Mine, Australia.*

Surface mining is less dangerous than underground mining; however, it does consume a huge amount of land. It requires the removal of a huge amount of the top layer of Earth and may lead to erosion, dust pollution, and loss of habitat. The mining process can also cause heavy metals to dissolve and seep into groundwater and surface water, which can contaminate drinking water sources and disrupt marine habitats.

Global reserves

Minerals "run out" when they become so rare and expensive that they can no longer be feasibly used. The amount of a mineral remaining in the ground and the difficulty of extraction will affect a mineral's price; as the price rises, another commodity will take the mineral's place.

An example of this process is indium, a crucial component of indium tin oxide (ITO), which is now used to make transparent conductive components used in solar panels and LCD screens. Indium is now six times more expensive than it was in 1994. This is because extraction has remained constant at around 600 tons per year, even though the development of new technology has increased demand.

The U.S. Geological Survey states *"indium's recent price volatility and various supply concerns associated with the metal have accelerated the development of ITO (indium tin oxide) substitutes. Antimony tin oxide coatings, which are deposited by an ink-jetting process, have been developed as an alternative to ITO."* Though recycling efforts have been made to harvest iridium from old phones and TV sets, it is estimated that there are now less than 14 years left of indium supplies.

In general, the price of most metals has increased by about 400% since 1994, which suggests that demand is outstripping supply. The demand for minerals mainly comes from China, which now consumes over half of the world's supply of various metals.

Relevant laws and treaties

Mining law in the United States is based on English common law. The landowner is the owner of all raw materials to unlimited depth. However, the state retains the rights to phosphate, nitrate, potassium salts, asphalt, coal, oil shale, and sulfur. The Department of the Interior has the rights to sand and gravel.

The *General Mining Act* of 1872 is a U.S. federal law that authorizes and governs prospecting and mining for economically valuable minerals, such as gold, platinum, and silver, on federal public lands. This law organized the informal system of acquiring and protecting mining claims on public land formed by prospectors in California and Nevada from the late 1840s through the 1860s during the California Gold Rush. It opened federal land in the public domain for prospecting and mining.

The *Hardrock Mining and Reclamation Act* of 2007 would have permanently ended new patents for mining claims, imposed a royalty of 4% of gross revenues on existing mining extracting from unpatented mining claims, and placed an 8% royalty on new mining operations. Of that royalty money, 70% would have gone to a cleanup fund for past abandoned mining operations; the remaining 30% would have gone to the affected communities.

Opponents of the bill said that it would have given U.S. mining operations the highest effective tax rate in the world and that implementing further restrictions and imposing royalties would force even more of the domestic mining industry out of the country. The bill died at the end of the 110th Congress. A later version of the bill, introduced in 2009, died at the end of the 111th Congress.

Some conventions of the International Labour Organization that govern mining in the United States are the Medical Examination of Young Person's Convention (Underground Work) of 1965 and the Safety and Health in Mines Convention of 1995.

Figure 4.17. *The Farmington Mine disaster, West Virginia.*

According to historical data on mining disasters, mining accidents claimed thousands of lives a year from 1880 to 1910, with over 3,200 deaths occurring in 1907 alone.

The Monongah Mining Disaster of 1907 was the worst mining accident in U.S. history. A total of 362 people were killed in the underground mines as a result of an explosion that occurred on December 6 of that year in Monongah, West Virginia.

In the 1950s, the annual death rate decreased to about 450. During the 1990s, there were only about 90 deaths a year; however, that does not consider the thousands of people injured in mining accidents between 1991 and 1999.

One of the more recent mining accidents occurred in West Virginia on April 5, 2010, in the Upper Big Branch mine, where 29 miners died as the result of an underground explosion.

To provide for the safety of miners, the U.S. government established the U.S. Bureau of Mines in 1910. The U.S. Bureau of Mines investigates accidents, advises the mining industry on what are the best safety practices, and teaches courses in accident prevention for miners, so they know how to react in the case of a mishap.

The U.S. government passed the *Federal Coal Mine Health and Safety Acts* in 1969 and 1977, which require multiple annual inspections of all mines.

Notes

Fishing

Fishing involves catching not only fish but other aquatic animals, such as echinoderms, crustaceans, mollusks, and cephalopods. Fishing does not refer to the act of catching aquatic mammals, such as whales. Fishing can be done for commercial or recreational purposes.

Figure 4.18. *Alaskan fishermen, 1927.*

Commercial fishing is done to sell the catch either through a previously agreed-upon contract, in a marketplace, or to an individual. In recreational fishing, fish are caught for personal consumption or sport and are not sold in markets. The fishing industry is growing rapidly throughout the world. It has also led to the growth of imports and exports.

Different varieties of fish are found in different regions of the world. The U.S. has a solid place in the fishery business, as it has an abundance of fishing and wildlife opportunities and millions of people who earn their livelihood from fishing. The National

Wildlife Refuge System is managed by the U.S. Fish and Wildlife Services and is responsible for protecting wildlife and aquatic animals.

Fishing Techniques

There are many methods of large and small-scale commercial fishing. Listed below are the most popular commercial fishing methods.

Pelagic Trawling: This method involves dragging a cone-shaped net through the water to catch fish in an area where they are feeding. This method does not have a large impact on the environment if it is conducted responsibly. Examples of the deleterious environmental effect of pelagic trawling can be seen in New Zealand and Australia.

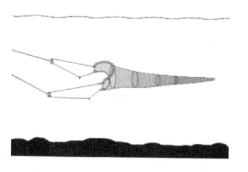

Figure 4.19. *Pelagic trawling*

Bottom Trawling: This is a method used to catch fish that live and feed on or near the seabed. This method is often referred to as demersal trawling, as the nets are dragged through the area of the sea known as the demersal zone. Species caught by this method include cod, haddock, sole, and whiting.

Otter Trawling: This method (shown below) involves a wide-mouthed net dragged through the water; the mouth of the net is kept open via the movement of the boat. This method disorients the fish and causes them to swim into the path of the net. Once this occurs, they cannot escape, even if they attempt to swim away; the fish find themselves stuck in the cod end of the net, pinned by the force of the water rushing past them.

Nets used for this method have holes of various sizes in the mesh, depending on the type of fish they are attempting to catch. Often the mesh is wide enough to allow younger, smaller fish to escape.

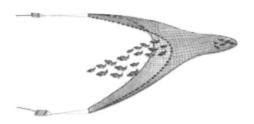

Figure 4.20. *Otter trawling*

Beam Trawling: This method is used to catch fish that burrow themselves under the seabed. A chain on the front of the net, which itself is lowered from a beam, strikes the ocean floor to scare the fish out from the safety of the seabed. When they emerge, they are swept up by the net.

Beam trawling is an extremely destructive form of fishing; it can have a deleterious effect on many aspects of the ecosystem. It destroys large swathes of the seabed that many species, not just fish, call home. According to estimates, for every pound of fish caught by beam trawling that finds its way to market, 16 pounds of marine life have been killed.

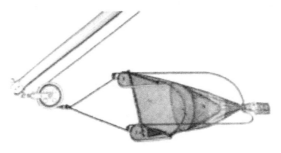

Figure 4.21. *Beam trawling*

Pulse trawling is another version of beam trawling, using an electric current that travels through the nets and stuns fish out of the seabed, allowing them to be scooped into nets. Fishermen claim it is less damaging than a traditional beam trawl, but there have been reports of masses of dead fish and devastation to the marine environment caused by this method.

Pair Trawling: This method involves two boats pulling a large net. The nets used in this method can be massive, with the largest nets able to fit 10 Boeing 747 jets. This method is extremely profitable, as the ratio of fish yield to cost expenditure, is great. However, due to the damaging effects pair trawling has on the environment, it is banned in many places around the world, though not in the United States.

Figure 4.22. *Pair trawling*

Purse Seining: This is a method used to collect fish that gather in schools such as herring, mackerel, sardines, and tuna. Purse seining uses a large net, open at the bottom, which is pulled across the water. When the target fish are within the radius of the net, the bottom is closed, trapping the fish. Then the net is pulled on board the boat. This method is not harmful to the environment if it is done sustainably; however, the scale on which this method is used is often unsustainable. Often, the nets are thousands of square meters, and hundreds of tons of fish can be captured at one time. This method often receives attention from the media because it all too often leads to dolphins being killed.

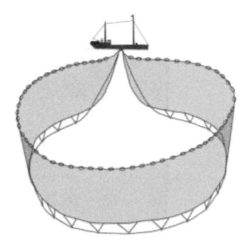

Figure 4.23. *Purse Seining*

Dredging: This method is used to catch shellfish that live on the seabed. Nets are dragged behind a boat on the seafloor, with "teeth" or water jets on the front of the nets. These are used to dislodge the shellfish from the seabed and force them into the nets.

This is extremely destructive; often, one pass of a boat using the dredging method will destroy an entire shellfish bed.

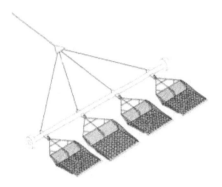

Figure 4.24. *Dredging*

Long-line Fishing: Long-lining is used commercially to catch species such as marlin, tuna, and swordfish. In this method, a long line with thousands of hooks with bait attached and left to drift behind the boat. This method is largely considered to be sustainable, in that it does not destroy the seabed ecosystem. However, endangered species from both the sea and air are often caught in long lines. It is difficult to control what species of animals get caught, as some boats may use lines 30 miles long.

Pots: This is a method used to catch lobsters and crabs. A cage (pot) is filled with bait and lowered to the seabed, and the mouth of the pot is left open. After a set period, the mouth closes, trapping those inside the pot. This method is considered sustainable, as it only targets the desired species.

Further, if undesired species get trapped in the pot, they can be thrown back once the pot is taken to the surface. Due to the low-impact nature of this type of fishing, both the intended and unintended species arrive on the deck of the ship alive.

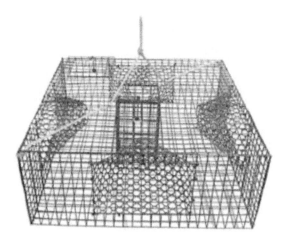

Figure 4.25. *Pots*

Overfishing

When the level of fishing taking place in an area is greater than the level of sustainable reproduction of the species in that area, a phenomenon known as *overfishing* occurs. Due to the level of competition in the fishing sector, there is a drive to catch the most fish in the shortest period; this results in overfishing, which is not conducive to a healthy ecosystem. It has a negative ecological impact on sea creatures and a negative socioeconomic effect on the local communities where such practices take place.

The WWF has made several efforts to protect fish and aquatic life. It has created several regulations and laws to maintain balance. This not only helps wildlife but also promotes sustainable fishing practices, which help fishermen. Due to its efforts, several national and international efforts have been made to control this activity.

In 2014, there was a declining rate of fishing in the U.S.; since 1997, fishing has been at an all-time low (NOAA). The U.S. is making respectable efforts to decrease the overfishing rate. It has made strides to restrict illegal fishing and overfishing. These efforts have helped a lot in safeguarding the marine environment throughout the world.

Aquaculture

The farming of fish in segregated or protected areas is known as *aquaculture*. It includes all the activities required to help in the proper growth of marine life, such as the rearing, breeding, and harvesting of marine plants and animals in all sorts of water bodies (oceans, ponds, lakes, rivers).

The farming of ornamental fish, crustaceans, mollusks, bait fish, sport fish, food fish, fish eggs, algae, and sea vegetables are all components of aquaculture. In this type of farming, extra care is given to marine life so that they can flourish.

Marine aquaculture produces oysters, mussels, salmon, shrimp, cod, and barramundi. Tilapia, trout, catfish, and bass are raised in freshwater aquaculture. The U.S. is very keen on both varieties of aquaculture.

Figure 4.26. *Aquaculture in Luoyuan Bay, Fuzhou, China.*

Relevant laws and treaties

The basis for all commercial fishing laws and policies in the United States is established on a federal level and enforced by the *U.S. Fish and Wildlife Service*. The federal government also establishes how the nation interacts with other fishing nations. Individual states dictate laws and policies within their waters, which extend three miles offshore. States must, however, abide by the minimum guidelines set forth by federal regulations.

The *Magnuson-Stevens Fishery Conservation and Management Act* of 1976 is the defining federal law for fisheries management. It defined federal waters, restricting foreign countries from fishing close to U.S. shores. It also established eight regional fishery councils, comprised of state and federal representatives and members of the fishing industry.

The councils are under the jurisdiction of the federal *National Oceanic and Atmospheric Administration* (NOAA), the agency responsible for regulation and enforcement. Each council develops fishery management plans for their specific area.

The *Sustainable Fisheries Act* was passed in 1996 as an amendment to the above act. It was designed to conserve fish stocks and restore overfished populations, minimize bycatch, restore habitat and assist traditional small-scale fishermen.

The *Magnuson-Stevens Fishery Conservation and Management Reauthorization Act* of 2006 took significant conservation steps designed to end overfishing. Those steps include policies such as catch shares and sector allocation. It also called for an increase in global cooperation.

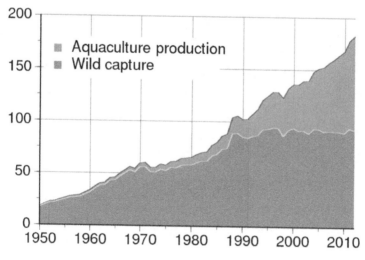

Figure 4.27. *Global total fish harvest.*

Global Economics

The global economy is the economy of the world, an international exchange of goods and services. Countries are ranked within the world economy by their nominal GDPs (Gross Domestic Product, the primary indicator of a country's economic performance) and estimates from financial and statistical institutions, which are calculated at market or government official exchange rates.

Nominal GDP does not take into account differences in the cost of living within different countries, and the outcomes can vary widely from one year to another based on fluctuations in the exchange rates of a given country's currency. Such fluctuations may change a country's ranking, even though they often make little or no difference in the standard of living of its population.

The advanced economies of various countries worldwide contribute to the advancement of the world economy. The world economy is valued or judged typically in monetary terms. Even the trade of illicit services or goods generates the economic growth rate of a country under the global economy. Within this sector, market valuations in local currency are converted to a single monetary unit using the notion of purchasing power, the method used for evaluating worldwide economic activity regarding U.S. dollars or euros.

The global economy is inexorably tied to the environment. For example, the annual value of trade in oceanic fisheries is valued at $5.9 billion. However, catch rates are in continuous decline, and almost 75% of the world's fish stocks have already been fished up to or beyond their sustainable limit.

The global value of plant-derived pharmaceutical products is more than $500 billion in industrialized countries. Of the medicines currently available, 40-50% are derived from natural products; for oncology and anti-infection medicines, the total is closer to 70-80%.

The global economy is steadily growing and using up natural assets at a terrific speed. The rate of loss of species and habitats is proceeding relentlessly at a speed many times higher than natural processes.

Figure 4.28. *Gulf Coast wetland draining, and development contributed to the widespread flooding brought by Hurricane Katrina.*

However, markets fail to capture most ecosystem service values. Existing prices only reflect — at best — the share of the total value that relates to provisioning services like food, fuel, or water, and their prices may be distorted. Even these services often bypass markets when carried out as part of the community management of shared resources. The values of other ecosystem services are generally not reflected in markets.

This is because many ecosystem services are *public goods* (i.e., *common goods*); they are often open access with no limits on their consumption. Private and public decisions affecting biodiversity rarely consider benefits beyond the immediate geographical area.

Benefits that will be felt over the long-term (e.g., from climate regulation) are frequently ignored. This systematic under-valuation of ecosystem services and the failure to install appropriate controlling mechanisms is one of the main causes underlying today's biodiversity crisis. Values that are not overtly part of a financial equation are too often ignored.

Globalization

Globalization is the process of international integration and assimilation arising from the interchange of world views, products, ideas, and other aspects of culture. Globalization can be defined as an international network of economic systems.

The origin of globalization predates the European Renaissance and the discovery of the New World, though today's economic scholars place it within the twentieth century. It began on a large scale only in the late nineteenth century. Globalization is a modern term that emerged in the 1970s, emanating from scholars, journalists, editors, and librarians.

In 2000, the *International Monetary Fund* (IMF) denoted four aspects of globalization: trade and transactions, capital and investments, migration and diffusion of knowledge.

Globalization refers to a wide variety of actions, including building dams or linking roads from one country to another, technological innovation (e.g., initially telephones, then mobile phones), the advancement of the internet in day-to-day transportation of goods and people, automotive innovations and the progression of technology that has led to global warming, the greenhouse effect and the depletion of the ozone layer.

The first great wave of Early Modern migration involved the forced transatlantic movement of 9-12 million slaves during the early to mid-19th century, a major movement of people comparable to the regional Arabic slave trade and the European migration to the New World during the Early Modern Period.

Today, most nations can only react to globalization. However, the United States, as the hegemon of the global economy, can control or regulate the pace and character of the globalization process itself. Globalization has greatly affected the United States and its citizens.

First, globalization has spread American influence throughout the world.

Second, it has opened more markets for the United States, which allows American companies to export and import various goods and services on a large scale.

World Bank

The *World Bank* is a financial organization that gives loans to developing or underdeveloped countries for purchasing capital and for overall economic development. It is a member of the United Nations Development Group and consists of two institutions: The *International Bank for Reconstruction and Development* (IBRD) and *The International Development Association* (IDA). The formal objective of the World Bank is to erode poverty. It promotes foreign investment and international trade across the globe. The World Bank was created alongside the *International Monetary Fund* (IMF) in 1944 at the Bretton Woods Conference. Many nations attended the conference, but the United States and the United Kingdom had the most influence because they were the most powerful countries present. The World Bank's headquarters are in Washington, D.C., and its president is traditionally an American.

The first country to receive a loan from the World Bank was France, though because the World Bank did not have much money initially, France only received half the amount of money requested, with strict conditions. The U.S. State Department asked the French government to remove its communist members; within hours of their compliance, France's loan was approved by the World Bank.

Gradually, the World Bank started facing competition when most European countries started getting loans from other sources; after that, the World Bank started focusing on non-European countries. From 1968 to 1980, the Bank focused on meeting the essential needs of developing countries. The size and number of loans increased since the focus shifted from things like improving infrastructure to small-scale ventures, such as creating social services. Early in 1989, in response to harsh criticism from many groups, the Bank started offering loans to environmental groups and non-government organizations (NGOs) to compensate for the effects of its past development policies.

Per the Montreal Protocols, the World Bank worked to control the use of 95% of the ozone-depleting chemicals by 2015. Since then, the Bank has put additional policies into effect to protect the environment while facilitating development. For instance, in 1991 the Bank claimed that to protect against deforestation, it would not finance any commercial logging or infrastructure projects that harm the environment.

To promote global public goods, the World Bank tries to control communicable diseases, such as malaria, by delivering vaccines to several parts of the world. In 2000, the Bank announced a "war on AIDS," and in 2011, it joined the Stop Tuberculosis Partnership.

Tragedy of the Commons

The idea of the tragedy of the commons was originally coined in 1833 by the English economist, William Forster Lloyd to describe a situation where individuals act in their self-interest without regard for the welfare of the whole group, which leads to the erosion of a common resource.

As an example of this principle, Lloyd used the grazing of cows on a common, where farmers each graze one cow. By grazing their cow on the common, the farmers lose nothing of their resources (i.e., the grass that would be consumed if the farmers put their one cow on their property). Each farmer could gain a competitive advantage by acquiring one more cow and grazing it on the commons, a perfectly rational act, as the action of adding one cow is not unsustainable. However, a problem arises if each of the farmers, acting out of individual self-interest, were to add one cow to the commons. This would become unsustainable for that pasture.

This notion can be applied to the environment as well. No one individual is responsible for a great deal of environmental degradation. However, when the actions of multiple individuals over time and across space are amalgamated, they have hurt the environment (e.g., soil erosion, ozone depletion, global warming, etc.).

In 1968, Garrett Hardin published an article called *"The Tragedy of the Commons"* in the journal *Science*. Hardin analyzed issues that were hard to solve through technical means, or *"a change only in the techniques of the natural sciences, demanding little or nothing in the way of change in human values or notions of morality."*

Hardin highlighted the increased human population, the massive use of Earth's natural resources and the welfare state. He proposed that if individuals believed or thought about their welfare alone, and not about society, then the number of children each family had would decrease because parents realized they would not be able to provide their children with ample resources without society's help.

The *"Tragedy of the Commons"* can also be applied to environmental problems like sustainability. The word "commons" in this sense connotes such natural resources as the atmosphere, oceans, rivers, fish stocks, energy, and any other shared resource not officially controlled or regulated.

Thus, the *"Tragedy of the Commons"* denotes the terrible and devastating plight of these resources due to human use and consumption without regard for future impact. The commons dilemma stands as a perfect example of a wide variety of resource issues today, such as water, forests, fish and non-renewable energy sources like oil and coal. The excessive mining for coal and other minerals is causing a threat to the ecological balance of the Earth.

Figure 4.29. *John Day Dam and fish ladder, Columbia River*

Conditions provoking the tragedy of the commons include the overfishing and destruction of the Grand Banks, the destruction of salmon runs on rivers that have been dammed (e.g., the Columbia River in the Northwest United States), the devastation of the sturgeon fishery in modern Russia and the limited water available in arid regions.

Relevant laws and treaties

The Earth's resources are shared by every individual living on the planet. Therefore, one of the foremost duties of world leaders should be to regulate the usage of natural resources by implementing different laws and treaties designed to protect the environment. Today, with over seven billion people on the planet, issues like air and water pollution, resource allocation, habitat destruction, and the burning of fossil fuels are prevalent.

Resource consumption and use have reached unforeseen levels, which has led to an increase in the number of environmental policies and regulations around the world. The environmental policy deals with human involvement with nature. This type of policy determines the regulation of resources and the reduction of pollution.

The regulations are generally designed so that they contribute to human welfare and the protection of natural resources. Since environmental issues are not restricted to the state, country, or regional boundaries drawn by humans, international cooperation is often needed to address them. Many international laws and conventions have been formulated to support this cause.

The *International Tropical Timber Agreement* was created in 1994 to *"promote the expansion and diversification of international trade in tropical timber from sustainably managed and legally harvested forests and to promote the sustainable management of tropical timber-producing forests."*

The Convention on Biological Diversity was created in 1992 with the objective of developing international strategies for the conservation, sharing, and sustainable use of biodiversity.

The *International Convention for the Prevention of Pollution from Ships* was enacted in 1973 to preserve the ocean from pollution caused by ships and other human activities.

The *International Convention for the Regulation of Whaling* is an environmental agreement signed in 1946 to provide for the proper conservation of whale stocks. It governs the commercial, scientific, and aboriginal subsistence whaling practices of 59 member nations.

The *International Commission for the Conservation of Atlantic Tunas* (ICCAT) is an intergovernmental organization responsible for the management and conservation of tuna and tuna-like species in the Atlantic Ocean and adjacent seas. The organization has been strongly criticized by scientists in the past for its repeated failure to conserve the sustainability of the tuna fishing by consistently supporting overfishing.

However, in recent years, ICCAT has adopted a strict recovery plan for Eastern Bluefin Tuna, which included strict monitoring, reporting, and control measures and reduction of the total allowable catches from 27,500 tons in 2007 to 13,400 tons in 2014.

Notes

Chapter 5

Energy Resources
& Consumption

Energy Concepts
- Energy forms
- Power
- Units
- Conversions
- Laws of Thermodynamics

Energy Consumption
- History
- Present global energy use
- Future energy needs

Fossil Fuel Resources and Use
- Formation of coal, oil and natural gas
- Extraction/purification methods
- World reserves and global demand
- Synfuels
- Environmental advantages/disadvantages of sources

Nuclear Energy
- Nuclear fission process
- Nuclear fuel
- Electricity production
- Nuclear reactor types
- Environmental advantages/disadvantages
- Safety issues

- Radiation and human health
- Radioactive wastes
- Nuclear fusion

Hydroelectric Power
- Dams
- Flood control
- Salmon
- Silting
- Other impacts

Energy Conservation
- Energy efficiency
- CAFE standards
- Hybrid electric vehicles
- Mass transit

Renewable Energy
- Solar energy
- Solar electricity
- Hydrogen fuel cells
- Biomass
- Wind energy
- Small-scale hydroelectric
- Ocean waves and tidal energy
- Geothermal
- Environmental advantages/disadvantages

Energy Concepts

Nearly all organisms rely on energy from the sun for survival. The primary producers in a food chain, such as plants, algae, and cyanobacteria, directly capture this solar energy and start the flow of energy through a system.

Energy is defined as the ability or capacity to perform work. Over the centuries, humans have developed an in-depth understanding of energy, including various methods to capture and utilize it for everyday activities.

Energy has several different forms, all of which measure the ability of a system or object to do work on another system or object.

Work is defined as a force applied over a distance.

Energy must be expended if work is done.

Energy forms

Energy is present in several forms, such as:

- *Thermal* (energy), or heat, can be measured by temperature. An object with a high temperature has more thermal energy than an object with a lower temperature.

  For example, when a container of hot soup is placed in a refrigerator, heat is removed from the soup as it cools, and the particles lose their thermal energy and begin to slow down. As a result of the slowing of the particles, the temperature of the soup decreases. A thermometer best measures thermal energy.

- *Chemical* (energy) is energy that is stored in the bonds of chemical compounds, such as molecules and atoms. When a chemical reaction occurs, and a bond is broken, the energy is often released as heat.

  For example, gasoline is a source of chemical energy used to move cars. As the gasoline burns, small "explosions" (breaking of bonds) occur which release heat that makes the pistons fire and the cars move forward.

- *Electrical* (energy) is the energy stored inside small particles called electrons. Electrical energy is generated when an electric field successfully moves electrons from one atom to another.

  For example, storm clouds build up large amounts of electrical energy, which is released as lightning when the clouds bump into one another.

- *Nuclear* (energy) is the energy in the nucleus of an atom, which holds the protons and neutrons together. Nuclear energy is released during nuclear fusion or fission.

  For example, nuclear power plants harness the process of fission of atoms to produce electricity.

- *Potential* (energy) is energy an object possesses due to its position or condition, rather than motion. For example, a raised weight, a battery, a coiled spring, and a boulder on top of a hill all have potential energy. This is also referred to as *stored energy*.

- *Kinetic* (energy), referred to as the *energy of motion*, is observable as the movement of an object or particles. The potential energy that is stored in an object becomes kinetic energy once the object begins moving. A falling weight, a coiled spring being released, and a boulder rolling down a hill are all examples of kinetic energy.

- *Mechanical* (energy) is the sum of both potential and kinetic energy. To determine the power generated from these energies, both motion and position must be considered. An object has mechanical energy if it is both in motion and in a vertical position above the ground.

  For example, a wrecking ball has potential mechanical energy when it is raised vertically from the ground.

- *Electromagnetic* (energy), also referred to as light, may be thought of as little packs of energy called photons. Photons are created when electrons jump from an area of high energy to an area with lower energy, releasing bursts of energy seen as light.

Many other forms of energy exist, but the ones listed above are the most commonly discussed in physical sciences.

Power

Power is defined as the rate at which work is done or the amount of energy consumed in a unit of time. The SI unit for power is the watt, which is equivalent to joules per second (J/s), and was named after James Watt, the man who perfected the steam engine in the 18th century.

Power and work may sometimes seem like similar concepts, but they are very different. *Work* refers to an action involving a force and a movement in the direction of that force.

For example, a force of 10 Newtons is pushing an object 2 meters; this action does 200 joules of work.

Power is the rate at which work is done, so if 200 joules of work is completed in one second, the power is 200 watts.

$$Work = Force \times Distance \qquad (N \cdot m)$$

$$Power = Work \:/\: Time \qquad (N \cdot m)/s$$

Units

Basic SI Units

Dimension	Basic Unit	Symbol
Time	Second	s
Length	Meter	m
Mass	Kilogram	kg
Temperature	Kelvin	K
Electric Current	Ampere	A

Derived SI Units

Dimension	Unit	Symbol
Power	Watt	$W = (J/s)$
Energy	Joule	$J = N \cdot m$
Energy Flux	Watt per square meter	W/m^2
Speed	Meter per second	m/s
Acceleration	Meter per second squared	m/s^2
Force	Newton	$N = kg \cdot m/s^2$
Area	Meter squared	m^2
Volume	Meter cubed	m^3
Pressure	Pascal	$Pa = N/m$
Voltage	Volt	$V = W/A$
Volume Flow	Cubic meter per second	m^3/s
Mass flow	Kilogram per second	kg/s
Calorific Value	Joule per kilogram	J/kg
Specific Heat	Joule per kilogram Kelvin	$J/(kg \cdot K)$

Conversions

Conversions of common non-standard international units, their symbol and their SI equivalents in joules (J)

Non-SI Unit for Energy	Symbol	Equivalence in SI-Units
Kilowatt-hour	kWh	3.60×10^6 J
Tera watt year	TWy	31.5×10^{18} J
British Thermal Unit	Btu	1.055×10^3 J
Horsepower-hour (metric)	hp·hr	2.646×10^6 J
Erg	erg	10^{-7} J
Foot pound force	ft·lbf	1.356 J
Kilogram-force meter	kgf·m	9.8 J
Calorie	cal	4.187 J

Laws of Thermodynamics

Thermodynamics is the study of the effects of energy, work, and heat on a system. Generally, thermodynamics only focuses on large-scale systems. There are three principal laws of thermodynamics, each of which defines the properties (e.g., temperature, energy, and entropy) that help comprehend and predict how a system operates.

Entropy is defined as the unavailability of the energy in a system to convert its thermal energy to mechanical work> Entropy is the degree of disorder in a system.

The Zeroth Law of Thermodynamics states that if two systems are in thermal equilibrium with a third system, they will be in thermal equilibrium with each other.

The First Law of Thermodynamics (the Law of Conservation of Energy) states that energy cannot be created or destroyed. The total energy put into a system is equal to the energy that comes out, though the energy may change forms.

For example, a plant receives solar energy from the sun, which gets converted to chemical energy in a process called photosynthesis. No energy was lost; it simply changed forms from solar to chemical energy.

The Second Law of Thermodynamics states that in any cyclic system, the entropy of that system will either remain the same or increase. In a closed system, the total useful energy at the beginning of a process will be greater than the useful energy at completion. This does not violate the First Law, as the energy that is "wasted" and no longer useful is emitted as thermal energy (heat).

The Third Law of Thermodynamics states that the entropy of a given system will approach a constant value as the temperature of that system approaches absolute zero. Typically, when a system is at absolute zero, its entropy is zero.

Notes

Energy Consumption

In the past hundred years, the human population has doubled twice. Human consumption of energy is doubling four times as fast as the population, and the number of automobiles in the world is doubling ten times as fast. Humankind began utilizing coal as a source of energy as early as 1300 to cook food and provide warmth, but it was not until the 1800s that more uses for coal were discovered. A major factor in this growing use of coal as an energy source was the Industrial Revolution of the late 17th and early 18th centuries.

It was not until the 1880s that coal was used to provide electricity for homes and factories. As coal and other fossil fuels came to be used to build factories and machines, power cars, and warm homes, the demand for fuel increased dramatically, as did drilling for resources. This high demand and dependence on non-renewable energy sources have worried some scientists and economists, as they recognize that stores of fossil fuels are limited. Some scientists estimate that if humans continue to use fossil fuels at the same rate as in the past several centuries, stores may run out shortly.

History

A monumental advancement in the understanding of energy was the mastery of fire. It can be used to prepare a meal and provide warmth. The burning of various biomasses led to the production of pottery and the refining of metals. In more recent history, fire was also a vital component of important inventions such as the steam engine, which converted chemical energy stored in coal and wood into kinetic energy. This remarkable invention was the mainspring of the Industrial Revolution, and it was utilized in the manufacture of machinery to power locomotives, ships, and eventually automobiles. Coal remained the largest source of fuel until the mid-20th century when it was replaced by petroleum.

Industrial Revolution

The Industrial Revolution was a period between approximately 1760 and 1840 in Great Britain when the transition from hand production methods to the use of machines occurred. The period was also marked by improved efficiency of water power, the increased use of steam power, the development of machine tools, and an increase in the number of factories. Textile factories employed most people in this era, and the standard of living began to increase for the first time in history.

Some of the most important technological advancements and innovations occurred in the United Kingdom, including the Watt steam engine. The steam engine was named after the man who perfected it in 1778, James Watt, and was made of iron and fueled by coal. His improvements allowed for greater efficiency of the steam engine, which was then used to power other machines and factories. The earliest mills were run by waterpower, but the introduction of the steam engine allowed factories to be built virtually anywhere, not just near water sources.

Exponential growth

Any time something is increasing by a percentage, it is increasing exponentially. To determine the amount of time it will take for something to double (e.g., the population of New York City), use the Rule of 70.

For example, New York City's population is said to be growing by 5% each year. Divide 70 by the percent increase to obtain the time (in years) it will take for the population to double.

70 / % growth = doubling time   70 / doubling time = % growth

70 / 5% = 14 years for the population of NYC to double

Common growth rates (% per year) and their respective doubling times in years, calculated using the Rule of 70

Growth Rate (% per year)	Approximate Doubling Time (years)
0.1	720
0.5	144
1.0	72
2.0	36
3.0	24
4.0	18
5.0	14
6.0	12
7.0	10
10.0	7

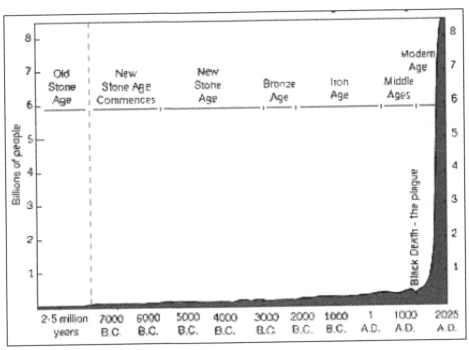

Figure 5.1. *The exponential growth of the human population throughout history.*

Energy crisis

The global reliance on fossil fuels and their relative scarcity have created major concerns about the possibility that there will be an energy crisis in the 21st century.

Oil, natural gas, and coal are all *non-renewable energy sources*, meaning that once they are used, they cannot be replaced. As worldwide industrialization continues, the demand for fuel sources skyrockets, and the supply continues to diminish.

The global consumption of oil each year is the equivalent of over 11 billion tons. Crude oil reserves are being used at an estimated rate of 4 billion tons per year. If this trend continues, the known oil deposits will be depleted by the year 2052.

Natural gas and oil will remain, but they will be depleted (as early as 2060) as they are used to fill the energy gap. Scientists estimate that if the world's population continues to grow exponentially, fossil fuels will be depleted even sooner than estimated.

Present global energy use

Modern-day global energy use can be characterized as being in a state of both slow transition and diversification. Specifically, as more nations and world leaders react to the reality of climate change and the growing dangers it possesses, investments and innovations are increasingly made to ween civilization off fossil fuels.

Large institutions, such as the *International Energy Agency* (IEA), the *European Environment Agency* (EEA) and the *U.S. Energy Information Administration* (EIA), track current consumption per source and periodically publish data. In 2014, fossil fuel use (e.g., petroleum, coal, natural gas) accounted for an estimated 78.3% (down 0.1% from 2013), nuclear power 2.5% (down 0.1% from 2013) and renewables 19.2% (up 0.2% from 2013).

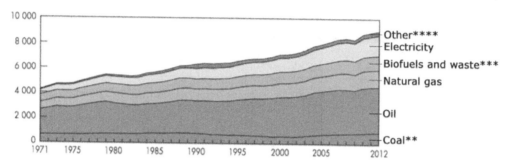

Figure 5.2. *World* total final consumption from 1971 to 2012 by fuel source in Mtoe (a megaton of oil equivalent). A toe is a unit of energy that is measured by the amount of energy released when one ton of crude oil is burned.*

*World includes international aviation and international marine bunkers.

**In this graph, peat and oil shale are aggregated with coal.

***Data for biofuels and waste final consumption have been estimated for some countries.

****Includes geothermal, solar, wind, heat, etc.

A shortlist of notable nations leading the charge in adopting renewable energy, as identified by Climate Council, include Sweden, Costa Rica, Nicaragua, Scotland, Germany, Uruguay, Denmark, China, Morocco, the U.S., and Kenya.

The continuing trend away from nuclear power is posited by experts to be a sustained reaction to previous nuclear disasters (e.g., Chernobyl, 1986; Three Mile Island, 1979; Fukushima Daiichi, 2011). Despite the strategic pivoting toward cleaner energy sources, fossil fuels are still overwhelmingly used worldwide.

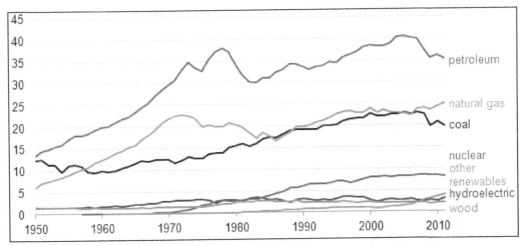

Figure 5.3. *U.S. primary energy consumption estimates by source, 1950 to 2011 (quadrillion Btu).*

Petroleum, natural gas, and coal have been the primary sources of energy for over 60 years. It was not until the 1970s and 1980s that other fuel sources such as nuclear and hydroelectric energy became widely used.

The renewables currently being adopted represent a wide range of sources: traditional biomass, bio-heat, ethanol, biodiesel, biopower generation, hydropower, wind, solar heating/cooling, solar PV, solar CSP, geothermal heat, geothermal electricity, and ocean power.

Unlike fossil fuels that exist in finite quantities, meaning that it is only a matter of time before they will be depleted, renewable energy comes from sources that are not depleted when used (e.g., wind or solar power).

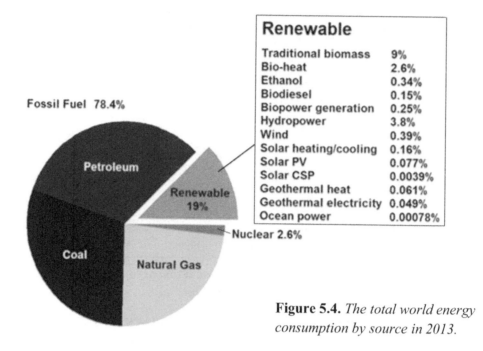

Figure 5.4. *The total world energy consumption by source in 2013.*

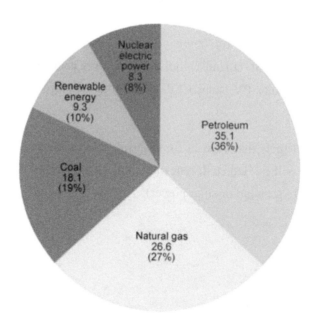

Figure 5.5. *Primary energy use by the source in the United States in 2013 (quadrillion Btu and percent total). Total use was 97.5 quadrillion Btu. Fossil fuels (coal, oil, and natural gas) provided nearly 80% of all energy used in the United States, and 10% came from renewable resources (wind, solar, and hydropower).*

Future energy needs

When scientists talk about the need for energy, they often refer to three-time frames; the short-term, the medium-term (to 2020) and the long-term (2050).

In the short-term scenario, the main issues are the prices of oil and whether the world can deal with the shock of the limited remaining energy.

In the medium-term scenario, emergent risks must be dealt with, and new sources of supply must be matured and utilized.

In 2050, the focus will be on resources beyond the current era of climate change and hydrocarbon concerns and instead on the ability and capacity to meet the energy needs of up to 9 billion people.

Despite the scenarios laid out for the future, analyses of pressing present needs, and what requires immediate action, scientists worry that there is too much speculation about future "what ifs" and not enough preventive action being taken today.

Obstacles faced in the ongoing transition to renewable energy efforts are many and often reflect systemic political and economic barriers.

A few leading obstacles that need to be navigated or overcome include:

new technologies competing with mature technologies (commercialization barriers);

existing subsidies and disproportionate tax burdens between renewables and fossil fuels/nuclear energy (price distortions);

market failures to value public benefits gained through renewables;

capital lack of access and split incentives (market barriers).

Notes

Fossil Fuel Resources and Use

Fossil fuels are defined as natural fuels that were formed in the past when prehistoric animals and plants died and were gradually buried by layers of rock and sediment. There are three main types of fossil fuels: coal, oil, and natural gas. Over millions of years, various combinations of the organic matter, the length of time buried, temperature, and pressure formed different types of fossil fuels.

Several countries across the globe produce oil and other products to meet the world's demand. Natural gas, petroleum, and oil (all fossil fuels) represent approximately 80% of all energy sources utilized in the United States.

At present, fossil fuels are extracted from the ground by drilling and used as a source of energy for transportation, machinery, heating, and electricity. The United States both produces and imports oil. For the first time in nearly 20 years, the United States is producing more barrels of oil domestically than are being imported from other countries (see Figure 5.4). According to the *Energy Information Administration* (EIA), fossil fuels meet around 82% of the U.S. energy demand (instituteforenergyresearch.org).

Formation of coal, oil and natural gas

Fossil fuels, such as coal, oil, and natural gas, were all once-living organisms. Fossil fuels were formed by prehistoric plants and animals that lived several hundreds of millions of years ago, decomposed and became buried under layers of sand, rocks, and mud. Some areas were covered by thousands of feet of earth, while others were flooded with waters that are now our oceans.

Different types of fuels were formed based on several factors: the combination of animal and plant remains, the length of time they were buried, and the pressure and temperature of the particles while decomposing.

For example, coal deposits were formed from the remains of dead trees, ferns, and various other plants that lived up to 400 million years ago. In some areas of the eastern United States, coal was formed in swamps that were covered by seawater containing large amounts of sulfur. As the seas dried up, sulfur deposits remained in the coal. Scientists are trying to come up with methods to extract this sulfur, which can become an air pollutant when the coal is burned and released into the atmosphere. Oil and natural gas were formed from organisms that lived in water. Over time, they decomposed and became buried in silt and sediment.

Coal is used primarily in the United States as an energy source to generate approximately 50% of our nation's energy. America has more coal than any other fossil fuel source and more coal reserves than any other country in the world.

Extraction & purification methods

Before oil and natural gas can be extracted, the reserves must be located. *Geophones* are sensitive sound receivers that are set out in several lines on the ground. Scientists will then create small explosions or vibrations on the surface. Sound waves travel through various mediums at different speeds, so when the geophones receive the reflected sounds, scientists can tell what material they bounced from under the surface. The sound waves travel slowest through gas, faster through liquids, and the fastest through solids.

If the material seems to contain either oil or natural gas, drilling will begin. To reach the material deep in the Earth's surface, rotary drilling rigs are used. First, a large metal pipe containing a metal drilling bit in the center that rotates at very high speeds is pushed into the Earth's surface, creating a hole that is lined with steel pipe casing that allows gas or oil to be pumped up to the surface. Over the years, a few new methods have emerged to allow greater efficiency during drilling and pumping.

Hydraulic fracturing, often abbreviated as *fracking*, creates small cracks in the ground called fissures, which allow gas and oil to flow out of solid shale rock.

Another new method is *directional drilling*, which allows a hole to be drilled straight into the earth but curves once it reaches deep underground. Often, oil is found below the ocean floor. Large platforms are constructed out at sea, which anchors the drill and containers for the oil.

Coal, a solid fossil fuel, is extracted from the ground through the surface and by underground mining. If the coal lies within approximately 60 meters (roughly 197 feet) of the surface, it will be extracted using various methods of surface mining. *Surface mining* uses large machines or explosives to remove the soil, animal remains, and plants that sit on top of the coal seam. It can also include the removal of mountaintops and open-pit mining. The extracted coal is then moved to processing plants located near the mine. Coal has various processing requirements based on the type and the location where it was mined. Processing plants clean the coal to remove contaminants and impurities, such as excess sulfur and mercury. Some plants use large amounts of water to wash the coal, which helps with removing impurities.

After processing, coal is transported by railroad or barge to the area where it will be used. Some coal is taken to power plants where it is crushed and used to make electricity. The wastewater from coal-processing can be dumped into underground mines or retention ponds because it degrades any water source it is added to. Retention ponds and other areas where this wastewater is dumped may leak or break, allowing for contamination of surrounding water sources.

Methane is the main compound found in natural gas that is utilized, but the gases pulled from deposits underground contain a variety of compounds. Once it has been extracted, the gas goes through initial processing on site. It is then transported by a pipeline to a processing plant to separate the methane from the other compounds, which include butane, propane, water vapor, hydrogen sulfide, and carbon dioxide. Some may be extracted for use in other industries, but it first must be purified and treated. For transportation to remote areas, the purified gas is transported as liquid natural gas via tankers. For more local transport, the gas will remain in pipelines.

World reserves and global demand

Fossil fuels are currently meeting around 80% of the global demand for energy. According to the 2013 Executive Summary of the International Energy Agency, even if countries make vast changes to deal with climate change now, their projections indicate the global energy demand in 2035 will increase by 40%, with fossil fuels still being the major source at 75%.

Economists believe this spike in demand will result from the increasing industrialization of countries such as India and China. Though fossil fuels will continue to meet the world's energy needs for the foreseeable future, there is growing concern about whether there will be enough resources to meet the continuing demand.

Fossil fuels around the world are currently abundant, but most are resources, not reserves. The key to considering our future energy needs is converting current resources into reserves.

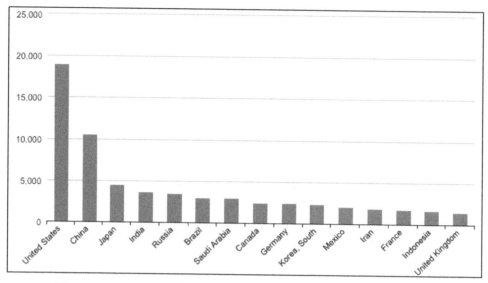

Figure 5.6. *Global consumption of petroleum by country in 2013 (in thousands of barrels per day).*

Currently, proven reserves of oil are estimated to be about 1.3 trillion barrels, with remaining resources (estimated to be recoverable from the ground) at about 2.7 trillion barrels. The *Organization of the Petroleum Exporting Countries* (OPEC) and other organizations ascertain that 101 nations still possess notable amounts of oil reserves, the top four being Venezuela, Saudi Arabia, Canada, and Iran.

As of 2014, the largest producers of crude oil were: The United States (13.9 million barrels per day), followed by Saudi Arabia (11.6 million barrels per day), Russia (at 10.8 million barrels per day).

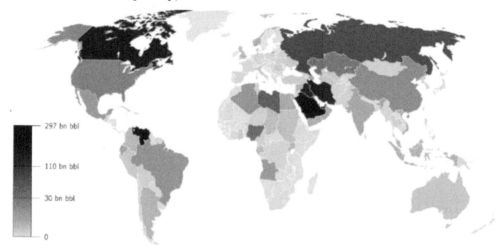

Figure 5.7. *World Oil Reserves, 2013.*

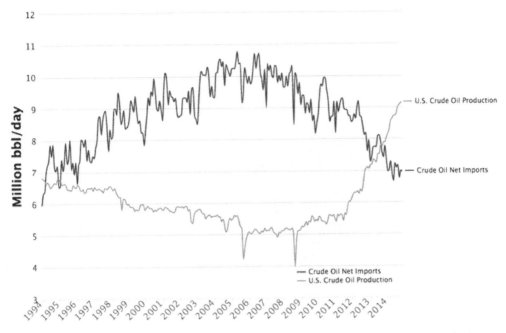

Figure 5.8. *U.S. Oil production and imports (in millions of barrels per day).*

Proven reserves of gas are estimated at 220 trillion cubic meters, or about the equivalent of 1.4 trillion barrels of oil and remaining recoverable resources at about 3.2 trillion barrels. The largest producer of natural gas is the U.S. (at 728 billion 200 million cubic meters (cu m) in 2014), followed by Russia (at 578 billion 700 million cu m in 2014); estimates put worldwide production for 2014 at roughly 3,460.6 trillion metric tons.

British Petroleum (BP), a leading energy company, states that 102 nations possess an exploitable amount of natural gas reserves, the top four being Russia, Iran, Qatar, and Turkmenistan.

Coal reserves are high, with proven amounts close to 730 gigatons (approximately 3.6 trillion barrels), with reserves estimated around 18 trillion barrels. The largest coal producer is China (at 3,874 million tons in 2014), followed by the U.S. (at 906.9 million tons in 2014); estimates put worldwide production for 2014 at roughly 8,164.9 million tons. More than 75 nations have significant coal deposits, the top four being the U.S., Russia, China, and Australia.

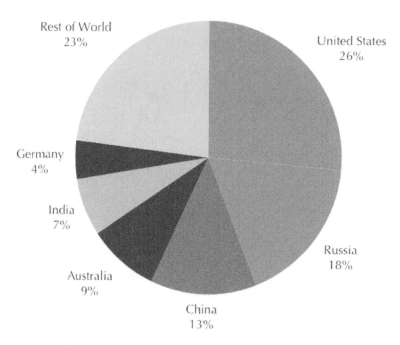

Figure 5.9. *Recoverable Coal Reserves by Country, 2011.*

Synfuels

Synthetic fuel, or synfuel, is a liquid or sometimes gaseous fuel obtained from syngas. *Syngas* is a mixture of hydrogen and carbon monoxide, and it is derived from the gasification of coal or biomass. Synthetic fuel is any liquid fuel obtained from natural gas or coal. Some scientists have started looking towards non-food crops to make synfuel. Since plants utilize carbon dioxide in photosynthesis, using them to create fuel would also cut down drastically on CO_2 emissions (nearly 50% as some scientists estimate).

Algae are very promising as a future biofuel, as it grows incredibly fast and is readily available. Though many believe that using synthetic fuel could meet our future energy needs, it would not be an easy or quick process.

For example, building a processing plant for converting algae into biofuel requires a large capital investment, and thousands of plants would have to be built across the nation for it to be a large supplier of fuel.

Some scientists, notably Christodoulos Floudas, a chemical and biological engineering professor at Princeton, estimate that it would take the United States approximately 30 to 40 years to fully adopt a synthetic fuel system, and it would cost upwards of $1.1 trillion for the entire system. Though this price is steep, it would become profitable in only a few years and provide less dependence on fossil fuels.

With the rising prices of oil and the projected prices for the future, the availability of an equivalent volume of synthetic fuel would be a relief to the average American's financial situation.

Environmental advantages & disadvantages of sources

Advantages

- Fossil fuels are relatively easy to locate.

- The power plants in use around the globe are very efficient at using natural gas as an energy source.

- Power plants for fossil fuels can be constructed in practically any location around the globe if the fuel can be transported there.

- Fossil fuels can generate great amounts of electricity.

- They are highly stable compared to other fuel substances.

- It is easily transported through pipes.

- Fossil fuels are currently the cheapest type of fuel on the market.

Disadvantages

- Perhaps the greatest disadvantage of fossil fuels is pollution. When they are burned, they give off carbon dioxide, which causes something known as the *greenhouse effect*. This is when radiation from the Earth's atmosphere is trapped by greenhouse gases, allowing heat to remain inside and warm our planet. Without greenhouse gases, this phenomenon would not occur, and the planet would be cooler.

- High levels of carbon dioxide and other greenhouse gases that are emitted by the burning of fossil fuels will not only cause a greenhouse effect but also lead to *global warming*. This refers to the rise in average global temperature (taken near the Earth's surface). Global warming is accelerating patterns in climate change and leading to changes in weather patterns, the seasons and sea levels.

- Coal produces carbon dioxide when burned, as well as sulfur dioxide, which is known to cause acid rain.

- The use and transportation of crude oils may pose environmental hazards during transportation, such as oil spills from tankers or leaks at drilling stations out at sea. An explosion of the Deepwater Horizon rig on April 20, 2010, took the lives of 11 men and thousands of sea life in the Gulf region.

  Efforts continued for months to stop the flow of oil from the rig, and according to a United States Coast Guard report, it is estimated that a total of 4.9 million barrels (210 million gallons) of oil were lost before they capped the leak 89 days later.

- Fossil fuels are a non-renewable resource, and current projections of future energy needs and fuel supply estimate that the planet will run out of fossil fuel by approximately 2080.

Nuclear Energy

There is immense energy in the nucleus of an atom, which holds the nucleus itself together. Atoms make up all matter in the universe, from grass and rocks to pizza and soda; even our air is made up of these tiny microscopic particles.

Nuclear energy is defined as the energy stored in the nucleus (core of protons and neutrons) of an atom.

Nuclear binding energy is the energy that would be required to disassemble the nucleus into its parts (protons and neutrons). The release or absorption of nuclear energy occurs in nuclear reactions or radioactive decay.

Italian-American physicist Enrico Fermi (Noble prize in 1938) discovered the potential of splitting atoms (nuclear fission) in 1934. Eight years later, Fermi created the first controlled nuclear reaction, using control rods and uranium. Using Fermi's technology, the production and testing of a nuclear bomb were only a few years away.

Nuclear fission process

Nuclear fission is defined as a nuclear reaction in which a large, unstable atom splits and releases energy. The fission of heavy elements releases large amounts of energy, which is an exothermic reaction. It occurs in heavier elements when the electromagnetic force, which pulls atoms apart, dominates the nuclear force, which holds an atom together.

A fission reaction is initiated when a high-energy neutron collides with an atom, producing an unstable isotope that undergoes fission. This fission process will produce additional neutrons, which can go on to initiate fission with other nuclei.

Nuclear fuel

The most commonly used fissile nuclear fuels are plutonium-239 (^{239}Pu) and uranium-235 (^{235}U).

The nuclear fuel cycle is composed of several steps, which include mining, refining, purifying, using, and disposing of nuclear fuel.

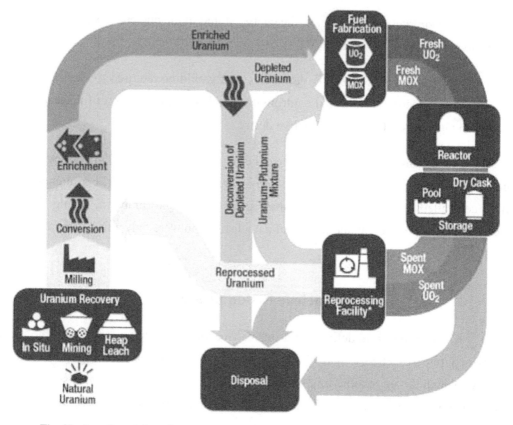

The Nuclear Regulatory Commission (NRC) has no regulatory role in mining uranium.

Figure 5.10. *The nuclear fuel cycle, showing the steps of uranium recovery, enrichment, utilization, and disposal of uranium.*

As a precursor to the nuclear fuel cycle, *uranium recovery* is the process of mining uranium ore from the Earth and milling or concentrating that ore. This recovery process produces a product called "*yellowcake*," which gets converted into uranium hexafluoride (UF_6). This compound undergoes enrichment, a process used to increase the concentration of uranium in UF_6. Reprocessing of the *Mixed Oxide Fuel* (MOX) is not practiced in the U.S.

The yellowcake is then moved to a fuel cycle facility where it is transformed into fuel. This nuclear fuel product will be used in reactors to produce nuclear power, which generates electricity.

Electricity production

Two types of power reactors, pressurized water, and boiling water are used mainly for producing electricity.

Pressurized Water Reactors (PWRs) have a four-step process to produce electricity, which the *Nuclear Regulatory Commission* (NRC) is as follows:

1. The core inside the reactor vessel creates heat;

2. Pressurized water in the primary coolant loop carries the heat to the steam generator;

3. Inside the steam generator, heat from the primary coolant loop vaporizes the water in a secondary loop, producing steam; and

4. The steam line directs the steam to the main turbine, causing it to turn the turbine generator, which produces electricity.

The *boiling-water reactor* (BWR) is the second type of power reactor:

1. The core inside the reactor vessel creates heat;

2. A steam-water mixture is produced when very pure water (reactor coolant) moves upward through the core, absorbing heat;

3. The steam-water mixture leaves the top of the core and enters the two stages of moisture separation where water droplets are removed before the steam is allowed to enter the steam line; and

4. The steam line directs the steam to the main turbine, causing it to turn the turbine generator, which produces electricity (www.nrc.gov/reactors/bwrs.html).

Nuclear reactor types

The United States *Nuclear Regulatory Commission* (NRC) regulates both power reactors and research and test reactors. Power reactors are those that generate electricity, and there are several types. Only two are in commercial operation in the United States: pressurized-water reactors and boiling-water reactors.

The NRC also regulates research and test reactors, many of which are used for training and research, testing of materials, and the production of radioisotopes for industry and medicine fields. These are much smaller than power reactors, and many can be found on university campuses around the world. They operate at lower temperatures and require less fuel.

The fission products that build up during fuel consumption are less than those seen in power reactors. Much like power reactors, these testing facilities require cooling, but only the high-powered test reactors need to use forced cooling, whereas all power reactors use forced cooling. There are around 240 research and test reactors in operation in 56 countries.

Environmental advantages and disadvantages

There is no doubt that the potential of nuclear energy is immense; the 104 commercial power plants in the United States produce about 20% of all electricity for the nation. However, there are definite downsides. The following outlines the positives of utilizing nuclear energy as a source of electricity in the U.S.:

Low costs: While the construction of power plants has a high cost, as does the enriching of uranium, it is still cheaper than generating electricity from natural gas, oil, and coal.

Base Load Energy: Nuclear power plants provide a good, stable base of energy. The electricity production from these sites can be lowered when wind turbines and solar panels are available.

Thorium: Yearly consumption reports show enough uranium for an additional 80 years. However, there is another product being used in China, Russia, and India called Thorium, to fuel their reactors. Also, thorium is a "greener" alternative to uranium.

Low Pollution: The environmental effects of nuclear power are low compared to other sources of energy, such as burning coal, oil, and natural gas.

High Energy Density: The amount of energy released during nuclear fission is approximately ten million times greater than the amount of energy released while burning an atom of oil or gas.

Sustainability: Currently, nuclear energy is not a renewable resource. However, some scientists believe that if they can develop a way to control the atomic fusion, the same type of reactions that occur to fuel the sun, we would have unlimited energy. Currently, many obstacles and challenges need to be addressed to make it a reality.

There are disadvantages to nuclear energy, which scientists are concerned about as serious threats to the environment and human life.

Radioactive waste: While nuclear power plants emit negligible amounts of carbon dioxide into the atmosphere, processes such as mining, enrichment, and waste management emit significant amounts, which are causes for concern.

Accidents: As seen in the Chernobyl incident of 1986 that emitted radiation after an explosion and fire in a nuclear power reactor, nuclear waste can have sudden, yet long-lasting effects. Estimates of the disaster concluded that it was responsible for 15,000 to 30,000 human deaths, and over 2.5 million Ukrainians are still experiencing health issues resulting from the high radiation levels. Further, some of the greatest effects of the disaster were felt as far away as Wales due to the toxic particles having found their way into the atmosphere and carried by the jet streams and clouds to many parts of Europe.

Safety issues

There have always been safety concerns about the use of nuclear energy, specifically regarding the safety hazards of the release of radioactive materials. During the manufacturing of power plants, design and operation were taken very seriously to decrease the likelihood of an accident and avoid health consequences if something were to occur. There have been three reactor accidents since nuclear power has been used: Three Mile Island (1979), Chernobyl (1986) and Fukushima (2011).

These are the only major accidents to occur, providing evidence that over the six decades that nuclear power has been utilized, it is a relatively safe means of generating electricity (compared to the over 7,500 deaths that occurred just *mining* coal for electricity in the past six decades, according to the U.S. Department of Labor). The risks associated with nuclear power are low and continue to decline.

Radiation and human health

Humans are constantly exposed to radiation from both natural and human-made sources. The risk associated with each source depends on how much radiation was absorbed, the time over which one was exposed to the dose and the exposure pathway of the radiation.

Radiation dose is defined as the amount of radiation that is absorbed by the human body, measured in *rems* (Roentgen equivalent man).

High doses of radiation can be extremely harmful and sometimes fatal. The damage that is caused by exposure to radiation can be determined by the part of the body that was exposed, the type of radiation, and the duration of the exposure. The effects may be seen promptly (several months after exposure) or be delayed and occur over several years. Delayed effects include many diseases, such as cancer.

There are two methods by which ionizing radiation can cause damage to the cellular structure: direct or indirect.

Direct radiation is when particles or rays strike and damage cellular components of a cell, such as its DNA.

Indirect radiation creates highly reactive atoms, or molecules, called free radicals that "grab" electrons from stable molecules and disrupt their function.

There are several forms of exposure, including:

Alpha particles, which are relatively slow and large. They cannot penetrate the skin but may be inhaled or ingested. Due to their large size and high energy, they may cause more damage than other forms of radiation.

Beta particles, which are about $1/2000^{th}$ the mass of alpha particles and can penetrate the skin. However, their small size reduces the potential to cause harm.

Gamma rays, which have very high energy and no charge or mass. They can pass through the very small spaces between cells without notice but may damage cellular structures they hit.

Radioactive waste

Radioactive waste contains radioactive materials, usually a byproduct of the generation of nuclear power, nuclear fission, and other nuclear technology (such as medicine and research).

Low-level waste (LLW) includes radioactively contaminated products, such as rags, medical tubes, tools, and protective clothing. Low-level waste disposal occurs at facilities licensed by either the Agreement States (37 states in agreement with the NRC) or the NRC itself.

High-level waste is a waste that is used nuclear reactor fuel. Currently, there is no single location used by the United States to dispose of high-level waste. A site at Yucca Mountain in Nevada was in development as a waste storage facility, but the government halted development in 2009. Many advocates want the repository to be utilized and are fighting for its completion.

High-level waste is now mostly stored on-site at nuclear weapons and nuclear power production facilities where the risk of leakage and build-up is high.

Nuclear fusion

Nuclear fusion is defined as a reaction in which two or more nuclei come close together and collide at very high speeds to join and form a new nucleus. During the process of nuclear fusion, energy is not conserved. Some of the matter that is present during the fusing of the nuclei is instead converted into photons (light energy). This release of light during fusion is what illuminates active stars in the night sky.

Energy is released when two nuclei with masses lower than iron-56 (^{56}Fe) undergo fusion, and energy is generally absorbed when two nuclei heavier than iron-56 undergo fusion (the opposite is true for the process of fission). This indicates that only lighter atoms such as helium and hydrogen are usually fusible, and only heavier elements such as plutonium and uranium are generally fissionable.

Notes

Hydroelectric Power

Hydroelectricity is defined as electricity that is produced by *hydropower*, the generation of electrical power by utilizing the gravitational force of flowing or falling water. It uses the energy of running water without reducing its quantity so that it can be categorized as a renewable energy source.

Hydroelectric power can be produced in various ways, from trapping it behind a dam and channeling a small amount to flow past a turbine at a time to small-scale hydroelectric systems that produce electricity for individuals on their land. Currently, China is the largest producer of hydroelectricity, followed by Canada, Brazil, and finally the United States. It is perhaps the oldest method of producing electricity known to man (e.g., the pre-Industrial Revolution mills).

Dams

Dams are often built on a large river that has a steep drop in elevation, and they store much water behind them. Near the bottom of the dam, there is an area that takes in water, which then travels down the penstock inside the dam itself. At the end of the penstock, there is a turbine that is propelled by the moving water. This turbine powers the generator above it that sits inside a powerhouse, which then sends electricity through power lines. The water that passed through the turbine then travels into an outflowing river.

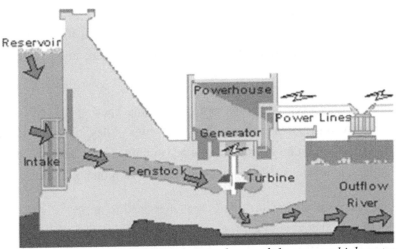

Figure 5.11. *A typical hydroelectric dam and the route which water takes to generate electricity.*

Flood control

Flood control is defined as all methods currently being used to lessen the damage of floodwaters. Many dams are partially or completely constructed to aid with controlling floodwaters. This design is slightly different from normal dam construction. The reservoir is often kept very low before an anticipated rainy or storm season, so there will be room for the water.

Dry dams are dams that serve the sole purpose of flood control. Mountain Morris Dam in New York and Seven Oaks Dam in California are two dry dams currently in use.

Salmon

While hydroelectric power is a very important source of energy, the operation of hydropower facilities is one of the many contributing factors to the decline of salmon species throughout the Pacific Northwest. Dams not only block the passage of salmon between their spawning (downstream) and rearing habitats (upstream) but also change water patterns in the river and raise water temperatures.

More than 55 percent of the spawning and rearing habitats once available to salmon are now permanently blocked by dams. Although dams were discovered to have a negative impact on salmon, few were ever removed.

A dam across the Wallowa River in Oregon was the first to be destroyed in response to declining salmon populations. The hope was that by opening the river, the salmon would once again thrive. Unfortunately, the salmon in the region had already adopted a migratory instinct around the construction of the dams and had become used to the lake-like environment.

Dams not only keep adult fish from swimming back up-river but also prevent juveniles from traveling downstream to lay their eggs. They must travel through the turbine, which spins at around 80 revolutions per minute. These high speeds usually kill or severely injure the juveniles, adding to the decline in salmon populations.

While many scientists worry that dams may lead to an irreversible effect on the salmon populations, others are much less concerned, thinking that hydroelectric power is more important than the fish.

Silting

Siltation is the pollution of water by fine particles such as silt or clay. It refers to an increased concentration of sediments suspended in the water, as well as an accumulation of particles on the bottom of a dam where they are not desired.

Though siltation is often considered a minor issue for most dams, it may reduce their life by decades. It reduces the amount of storage for rainwater, may damage the turbines, and can even impact the river downstream.

There are two main strategies currently in use for reducing the sediment that enters a reservoir: 1) preventing erosion, or 2) trapping the eroded sediment before it reaches the reservoir. Russia is currently experiencing issues with sediment in their dams. An efficient strategy was proposed to prevent it, in which upstream reservoirs would act solely as sediment-retention structures.

Other impacts

Contrary to natural gas and fuel, river water is not subject to price fluctuations. Hydroelectricity also contributes to the nation's stock of drinking water. The power plants collect rainwater in their reservoirs, which can be used for irrigation or consumption.

One of the advantages of hydroelectric power production and usage is that the hydroelectric cycle produces very small amounts of greenhouse gases. It is regarded as a green energy source, cutting down on the amount of pollution emitted into the atmosphere, as well as on acid rain and smog.

Notes

Energy Conservation

Energy conservation is defined as the act of reducing energy consumption by using less of that source of energy or choosing to go without that energy source altogether. For example, choosing to drive a car less is an example of trying to conserve energy. Some countries have decided to charge an energy or carbon tax to get their citizens to reduce the amount of energy they expend or the amount of carbon dioxide they release into the atmosphere (often called their *carbon footprint*).

Several industries have made efforts to reduce the amount of energy that is expended while using their product, as humans have become more aware and concerned about their ability to impact the environment. *Hybrid electric vehicles* (HEV) and *light-emitting diode* (LED) light bulbs are two products that aim to promote energy conservations.

There is an event called "Earth Hour," organized by the *World Wildlife Foundation* (WWF), to make the human population more aware of their impact on the planet Earth by having them turn off their non-essential lights for one hour out of the year.

The United Nations has attempted to address the issue of climate change through establishing international agreements; unfortunately, the adoption of such agreements to limit emissions has largely been rejected by the largest industrial polluting nations, such as the United States, China, and India.

Energy efficiency

Efficient energy use refers to using less energy for a constant task. For example, driving your car the same distance as usual, but choosing one that performs better on gas mileage is more energy efficient. When an appliance such as a computer or clothes dryer is replaced with a more energy-efficient model, the new appliances can provide the same service as before while using much less energy. This not only reduces the number of greenhouse gases being emitted into the atmosphere but reduces an electricity or energy bill.

Light-emitting diodes (LED) light bulbs provide 55-70 lumens per watt, as compared to a traditional incandescent light bulb that provides only 13-18 lumens per watt.

In the last few years, some housing and apartment complexes have made it mandatory for tenants to use LED lighting.

Entire homes have been made more energy-efficient by using wider lumber (which provides more room for insulation and makes the home virtually airtight to prevent air or moisture from penetrating the frame) and very high-efficiency water heaters and air conditioners.

Figure 5.12. *A "Zero Energy Home" located in New Paltz, New York. The average monthly energy bill for a Zero Energy home is $47. Building homes with very efficient energy devices, along with insulating extensively and using energy-saving windows, reduce the use of fossil fuels and carbon emissions.*

CAFE standards

The *Corporate Average Fuel Economy* (CAFE) standards are regulations that have been enacted by the United States Congress to improve the average fuel economy of cars, vans, and trucks that are sold in the U.S.

When studies emerged about the effects of car emissions on climate change, car manufacturers were encouraged to improve fuel efficiency and reduce carbon pollution. If manufacturers do not follow the regulations set by Congress, there are fines they must pay, some of which extend into millions of dollars.

As an incentive to become more energy-efficient, the *Environmental Protection Agency* (EPA) and the *National Highway Traffic Safety Administration* (NHTSA) list benefits that come along with the program of the CAFE.

Following the CAFE's program saves operators money by decreasing the cost of transporting freight, upwards of $170 billion. If a customer is happy with a car that is energy-efficient and knows it is decreasing their carbon footprint, it is assumed that sales will rise, and the car dealers' image enhanced by complying with regulations.

Hybrid electric vehicles

A *hybrid electric vehicle* (HEV) combines a conventional internal combustion engine with an electric propulsion system. The electric powertrain in the vehicle is intended to get better performance out of the vehicle or achieve better fuel economy.

Although the price of a hybrid electric vehicle is generally higher than that of conventional vehicles, the higher cost can be somewhat made up for by state incentives and fuel savings.

The most common type of HEV is the hybrid electric car, although hybrid electric buses and trucks are also in production. These vehicles use technologies such as regenerative brakes, which convert the kinetic energy from the motion of the vehicle into electrical energy to charge the battery. Other HEVs use their internal combustion engine to spin an electrical generator that charges the car's battery or generates electricity to power the electric motor.

Plug-in hybrid electric vehicles (PHEVs) use a battery to power an electric motor and another source of fuel, such as gasoline, to power a propulsion source such as an internal combustion engine. PHEVs have larger batteries than regular vehicles, which makes it possible to drive anywhere from 10 to 40+ miles on a fully charged battery.

All-electric vehicles (EVs) use only a battery to power the motor. The EPA classifies these as zero-emissions vehicles because they produce no exhaust. EV can typically drive around 100 miles on a single battery charge.

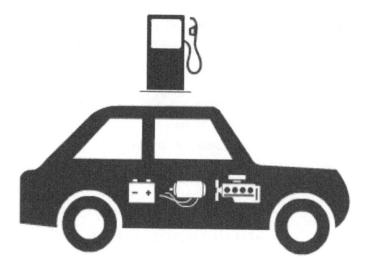

Figure 5.13. *Most hybrid electric vehicles have an internal combustion engine and an electric motor. These vehicles are powered by an alternative fuel or a conventional fuel, such as gasoline, and a battery, which is charged by regenerative braking.*

Mass transit

In 2007, the Intergovernmental Panel on Climate Change report cited studies that conclude that to limit global temperature rise to no more than 2 °C, emissions of greenhouse gases must be reduced by 40-50% from the 2000 levels by 2050.

Nearly 28% of all greenhouse gasses emitted are due to burning fuel for transportation; with an increased number of vehicles, the number of greenhouse gases being released will increase.

The use of public transportation reduces travel by individuals in their private vehicles. By choosing to ride mass transit vehicles such as buses, subways, and the metro, someone's carbon footprint can be significantly reduced.

Public transportation is estimated to save a household an average of $6,251 annually and to reduce carbon dioxide emissions by nearly 37 million metric tons every year.

Renewable Energy

Renewable energy is defined as energy that comes from resources that can be naturally replenished on a human time frame, as opposed to an earthly or universal one. These include energy sources such as sunlight, wind, waves, geothermal heat, rain, and tides. Also called clean energy, renewable energy swept across America in the past few decades as people became aware of the need for less dependence on fossil fuels and the harm they were causing. Creating new products that run on clean energy is a billion-dollar industry, but even so, only about 9% of all energy in the United States currently comes from renewable resources.

Solar energy

Solar energy is heat and light from the sun that can be harnessed for various purposes, most of which include the generation of electricity. It is the cleanest and most abundant renewable energy source that is currently available for use in the world.

The United States is considered the most advanced country for solar energy use, as it has the most modern technology to capture sunlight and harness it for heating water, generating electricity, and providing a light source for both indoor and outdoor use.

In 2011, the *International Energy Agency* (IEA) stated:

"The development of affordable, inexhaustible, and clean solar energy technologies will have huge longer-term benefits. It will increase countries' energy security through reliance on an indigenous, inexhaustible, and mostly import-independent resource, enhance sustainability, reduce pollution, lower the costs of mitigating global warming and keep fossil fuel prices lower than otherwise. These advantages are global. Hence the additional costs of the incentives for early deployment should be considered learning investments; they must be wisely spent and need to be widely shared."

The potential for solar energy to provide our energy needs has yet to be fully tapped. About 30% of solar radiation that strikes Earth's upper atmosphere is reflected in outer space. Clouds, water sources and land absorb the remaining radiation. The Earth's surface receives, on average, 164 Watts of solar energy per square meter.

Solar electricity

Solar power is the conversion of sunlight into electricity, and it is expected to become the largest source of energy for the planet by the year 2050. There are two types of systems that convert sunlight into electricity: *photovoltaics* (PV) and *concentrated solar power* (CSP). *Photovoltaics* involves the use of solar cells to convert solar light directly into electricity using the photoelectric effect (emission of electrons). This method is becoming increasingly popular using solar panels on houses and businesses to generate most or all their electricity needs. Concentrated solar power (CPS) uses a system of lenses or mirrors and tracking systems to concentrate an area of light into a small beam. This concentrated beam then produces heat that is used as a heat source.

Solar power is used in a wide array of industries, from agriculture to transportation. Agriculture and horticulture seek to optimize the amount of solar energy captured to provide optimal conditions for plant productivity. Greenhouses have been constructed with roofs made of solar panels, which convert solar light to heat. This allows for the year-round cultivation of plants in optimal light and heat conditions.

The development of a solar-powered car has been in the works since the 1980s. Some vehicles use solar panels to perform smaller electric jobs than powering the whole car, such as running the air conditioner, which reduces the amount of fuel consumption. A solar boat was constructed, which crossed the Pacific Ocean in 1996 solely on electricity generated from sunlight.

Hydrogen fuel cells

A *fuel cell* is a device that converts chemical energy from a fuel source into electricity. This is achieved through a chemical reaction of oxygen (or another oxidizing agent) with positively charged hydrogen ions.

Unlike batteries, fuel cells require a constant source of oxygen or fuel to keep the chemical reaction going. They can supply constant electricity if a fuel source is present and have relatively long lifetimes.

Fuel cells can be used for many applications, including backup power sources and transportation. They are more efficient than the standard combustion engine and have lower emissions because they only emit a byproduct of water. They are also very quiet during operation because they have no major moving parts. Due to their lack of moving parts, a fuel cell can achieve up to 99.9999% reliability under perfect conditions. They are often used as power sources for spacecraft, research stations, remote weather stations and various kinds of military machinery.

Biomass

Biomass power comes from plants such as corn, grasses, trees, and even algae. Biomass can either be used directly to produce heat, such as through burning wood or another plant-based biomass or indirectly after it has been converted to biofuel. To date, wood remains the largest source of biomass energy in use.

Biomass can be grown industrially to produce large quantities to be converted into biofuels such as ethanol (made from corn) and butanol (produced by fermentation of biomass by bacteria). Biofuels are generally classified into two major categories based on the source of the biomass they came from.

First-generation biofuels are derived from biomass such as sugarcane and cornstarch. Bioethanol is produced from the fermentation of the sugars and can be used as an additive to gasoline or to directly power a fuel cell.

Second-generation biofuels utilize non-food source biomass such as waste products from agriculture and municipalities. These lignocellulosic biomasses are not edible and are considered essentially non-valuable to the waste industries. Though it is preferred to create alternative fuel sources from non-food biomass, the lignocellulosic material is rigid and can be difficult to work with during the manufacturing process.

The most recent research in the field of biofuels is concentrated on *third-generation biofuels*, or fuels produced from algae or algae-derived biomass. Due to the speed at which algae grows, algae-based biofuels can be produced at up to 10 times the speed of other biomass-based fuel sources such as corn and soy.

The initial cost to create a plant capable of growing, harvesting and creating biofuel from algae is steep, but scientists estimate that they would pay for themselves in just a few years of use. Algae are a natural, renewable resource that can be harvested from the oceans and grown indefinitely (under the right conditions).

Wind energy

Wind Power comes from wind turbines or sails that move or rotate from flowing air passing over them. The wind turbine captures the natural wind in the Earth's atmosphere, converts it into mechanical energy, and later, into electricity. When the wind blows past a turbine, the blades capture the force and rotate.

The rotation causes an internal shaft to spin, which causes the gearbox to spin. This is connected to a generator that will produce electricity.

Wind turbines often stand together in a known windy area, such as on a large plain, a hilltop, and even out at sea. They are connected to a power grid so they can provide energy to the sources that need it.

There are three major types of wind power:

- *Utility-scale wind*, which uses turbines larger than 100 kilowatts. These turbines deliver electricity to the power grid, which is then routed to the end-user by a power systems operator;

- *Distributed* or *"small" wind*, which uses turbines smaller than 100 kilowatts. These deliver power directly to a house, small business or even a farm for their primary use; and

- *Offshore wind*, which uses turbines that sit on large bodies of water to capture the natural wind that blows across it (not yet utilized in the U.S. but under development).

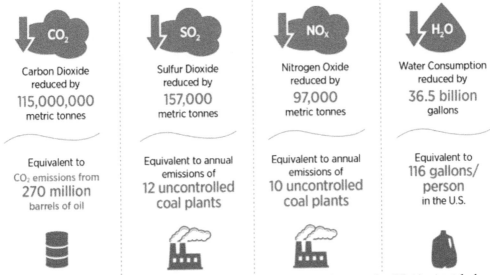

Figure 5.14. *Emissions and water savings calculated using the EPA's Avoided Emissions and Generation Tool (AVERT), due to the energy generated by wind power in 2013. Wind generation provides a range of environmental benefits. "Uncontrolled coal plants" are those with no emissions control technology. The United States ranks second in the world for installed wind capacity.*

Small-scale hydroelectric

Hydropower, or water power, is power utilized from the energy of fast-running or falling water. When most people imagine hydropower, they think of the Hoover Dam, a gigantic facility that harnesses the power of the entire Colorado River behind it. Hydropower facilities come in all forms and sizes; some are large, like the Hoover Dam, but others are small, taking advantage of small streams of running water.

For a typical micro-hydropower system that uses a small river to generate power, a portion of the river's water is diverted to a section on-land that includes a canal, forebay, a penstock, and a powerhouse. The water is delivered to a turbine or a waterwheel, which moves and spins a shaft. The motion of the shaft can be used for pumping water or for powering a generator.

A micro-hydropower system can be connected to an electrical grid or stand-alone. Stand-alone systems are used to power a home or small business.

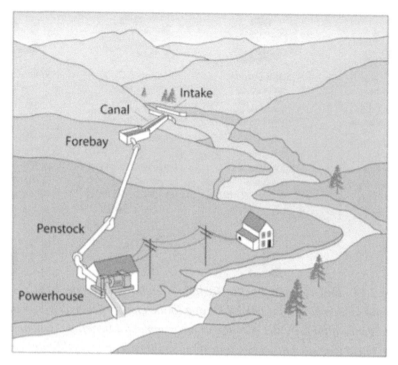

Figure 5.15. *An example of a small hydropower system on personal property. In this micro hydropower system, water is diverted into the penstock.*

Some generators can be placed directly into the stream. The system takes advantage of a water system that runs slightly downhill. The intake at the top allows water to travel down to the powerhouse, where the power of the water rotates a turbine. The turbine powers a generator, which produces electricity. This is an example of a stand-alone system.

Ocean waves and tidal energy

Tidal power is a form of hydropower that converts the energy from tidal waves into useful forms of energy, like electricity (also referred to as marine and hydrokinetic energy, MHK). Tidal energy is extremely promising as a future source of energy generation as tidal patterns are far more predictable than either wind patterns or the weather conditions that obstruct direct sunlight for the generation of solar energy.

Oceans cover approximately three-fourths of the Earth's surface and are one of the biggest potential renewable energy sources on the entire planet. The United States Energy Department estimates that the maximum amount of electricity that could be generated by harnessing the power of tidal waves is close to 1420 terawatt-hours, about one-third of the total annual electricity usage.

In addition to harnessing tidal power, energy from ocean waves can also be captured to generate electricity. Waves are caused by wind moving over the surface of ocean water. Most areas of the ocean around the world have predictable wind patterns, which provide an opportunity for harvesting energy through their constant movement.

Figure 5.16. *The Agucadoura Wave Farm off the shore of Portugal was one of the first of its kind. It was first tested and operated in 2008, just three miles offshore. This is just one of the four common types of devices used for capturing wave energy.*

Geothermal

Geothermal energy is energy in the form of heat from the Earth's surface. It can be found in deep wells in the ground and even in an average backyard. Many countries are already utilizing geothermal energy to reduce the number of greenhouse gases being released into the atmosphere by the burning of fossil fuels.

Under the Earth's crust, there is a layer of hot, flowing rock called magma. Heat is produced in this section by the decay of radioactive materials such as potassium (K) and uranium (U). The amount of energy contained within magma is thousands of times more than the energy in all the oil and natural gas in the world.

The areas with the highest underground temperatures are *hotspots*, occurring where tectonic plates meet. There is great potential in geothermal energy, as even the energy in the rock below the ground in our communities can control the heat in our homes.

Using current technologies such as the *Enhanced Geothermal Systems* (EGS), scientists are strategizing how they can capture this heat energy to produce electricity on a much larger scale than what is currently being used.

Environmental advantages & disadvantages

Hydropower is fueled by water, so it is considered a clean energy source as it does not produce the same level of carbon emissions seen with hydrocarbon-based sources of energy generation. Hydroelectric power is a domestic energy source so that each state can produce its energy without the reliance on fuel sources from outside the nation.

Hydropower plants can generate power immediately and can provide backup power during electrical outages. Micro-hydropower systems allow small communities or individual landowners to control and produce their electricity.

Fuel cells can be expensive, as platinum is a component of the system. They do not need to be recharged and do not become run-down easily. Fuel cells do, however, require a constant source of hydrogen to produce electricity. They are relatively durable, with stationary cells effective for roughly 40,000 operating hours.

Fuel cells also have low emissions, as they only produce water as a byproduct. This contributes zero pollution, in contrast to current energy sources such as fossil fuels, which release pollutants such as carbon dioxide and other greenhouse gases into the atmosphere.

Biofuels can be created from food and non-food sources, as well as biomass such as algae. They can be added to gasoline or used to power fuel cells. First and second-generation biofuels take longer to produce than algal biofuel, as the crops take longer to grow and harvest than algae.

However, the cost of creating enough stations for the nation to become reliably dependent on algae biofuel is very steep, like the growing, processing, and converting of algae to biofuel requires expensive equipment.

Research is currently underway to find an alternative method that is cheaper but equally efficient.

Geothermal energy has much potential for providing energy in the form of heat, which can be used to produce electricity. Though investments must be made to fund such large-scale projects, the future looks bright if we can capture the Earth's natural heat and use it as a source of energy.

Using geothermal energy would decrease the number of greenhouse gases being released, as well as slow down the rate at which fossil fuels are currently being used.

Over 25 billion barrels of hot water are being produced each year by gas and oil wells; the heat energy from those barrels alone could provide the United States with over 3 gigawatts of clean energy.

Notes

Chapter 6

Pollution

Pollution Types

- Air pollution

- Noise pollution

- Water pollution

- Solid waste

Impacts on the Environment and Human Health

- Hazards to human health

- Hazardous chemicals in the environment

Economic Impacts

- Cost-benefit analysis

- Externalities

- Marginal costs

- Sustainability

Pollution Types

Pollution is the contamination of an environment through the introduction of substances which hurt the natural state of the environment. These contaminants are referred to as *pollutants*.

There are seven types of pollution: land pollution, air pollution, water pollution, noise pollution, light pollution, thermal pollution, and visual pollution.

Air pollution

The Earth's atmosphere is made up of a mixture of complex gases that support life. The process by which harmful materials are introduced into the atmosphere, resulting in a threat to plant and animal health and natural ecosystems, is referred to as *air pollution*.

Air pollution has both natural and anthropogenic sources. Natural sources of air pollution include volcanic eruptions; this process releases a significant amount of pollutants such as carbon monoxide, sulfur dioxide, and particulate matter into the atmosphere, with adverse effects on life, property and the natural environment.

However, human activities are largely responsible for air pollution; the burning of fossil fuels from factories, motor vehicle exhaust, and cooking releases toxic levels of sulfur dioxide into the atmosphere.

Sources – primary and secondary

There are two types of pollutants; primary and secondary pollutants.

Primary air pollutants are pollutants emitted from a direct source, such as carbon monoxide from motor vehicle exhaust.

Secondary air pollutants are not directly emitted into the atmosphere. These pollutants are instead formed by the interaction of primary pollutants in the atmosphere to form an entirely different compound.

An example of this is ground-level ozone formed from the interaction hydrocarbons (C_nH_{2n+2}) and nitrogen oxide (NO_2) in the presence of sunlight.

Figure 6.1. *Exhaust from a diesel truck.*

Major air pollutants

One major primary air pollutant is *carbon monoxide* (CO), a colorless and odorless toxic gas formed from the incomplete combustion of fossil fuel. The main source of this gas is motor vehicle exhaust. It combines with hemoglobin in the body to produce carboxyhemoglobin, which reduces the amount of oxygen the blood can carry, reducing the oxygen supply to the body. This can result in seizures, coma, and death.

Sulfur dioxide (SO_2) is a poisonous gas that has a pungent scent. This gas is produced by volcanic activities, combustion of fossil fuels, industrial processes, and electric utilities. This gas has adverse effects on health because it aggravates the respiratory system and can cause asthma and lung cancer.

Sulfur dioxide is the primary pollutant that causes acid rain, which harms ecosystems. The release of this gas also contributes to global warming and can cause erratic weather patterns.

Nitrogen dioxide (NO_2) is a reddish-brown pollutant with a sharp, biting odor. NO_2 gas is emitted from high-temperature combustions of fossil fuel, particularly from motor vehicle exhaust, during thunderstorms by electric discharge, electric utilities, and industrial boilers.

Nitrogen dioxide gas is also a major contributor to the formation of ground-level-ozone, eutrophication, and acid rain. This pollutant harms aquatic and terrestrial life. Nitrogen dioxide also has an irritating effect on the lungs, resulting in respiratory problems in children.

Volatile organic compounds (VOC) are chemicals comprising hydrogen (H), carbon (C), and other elements in the form of solids, liquids, or gases that easily evaporate into the atmosphere. VOCs are categorized as methane (CH_4) or non-methane (NMVOC).

Methane is a greenhouse gas that plays a major role in global warming. It can be released from swamps or as by-products of livestock. NMVOCs such as toluene, xylene, and benzene are believed to be natural carcinogens. Sources of VOC include fossil fuel deposits, volcanic eruptions, and motor vehicle exhaust.

Many VOCs are common household products such as cleaning solvents, paint, and coatings. Other human-made sources of VOC include chlorofluorocarbons, which were once used in the manufacturing of cleaning products and refrigerants before being banned. These pollutants are hazardous to human health and the environment because they deplete the ozone layer.

Chlorofluorocarbons (CFCs) are a group of organic compounds that contains chlorine, carbon, and fluorine produced as a derivative of methane, ethane, and propane. CFCs are released from refrigerators, aerosol sprays, and solvents. This compound is a major contributor to ozone depletion. This depletion allows harmful ultraviolet rays to reach the Earth, resulting in damage to plants and causing skin cancers in animals.

Particulate matter (PM) are microscopic particles of solids and liquids suspended in the atmosphere that damage the environment. This includes a mixture of soot, dust, pollen, smoke, and water droplets of varied size and origin. Particulate matter may originate from a volcanic eruption, dust storms, combustion of fossil fuels and ocean spray. PM is believed to be the cause of many respiratory diseases, including lung cancer.

Secondary air pollutants include *acidic atmospheric pollutants,* which can be acid rain or any other precipitation that is acidic. *Acid rain* is caused by the reaction of rainfall to sulfur dioxide and nitrogen oxides to produce acids.

Sulfur dioxide and nitrogen oxides are released from power plants, electric utility plants, and the burning of fossil fuels. Dilute *sulfuric acid* (H_2SO_4) is formed when precipitation encounters sulfur dioxide in the air.

Carbon dioxide is the main greenhouse gas and is considered the main pollutant when emitted by airplanes, cars, factories, and the burning of fossil fuel. Carbon dioxide reacts with water to form carbonic acid.

Figure 6.2. *Industrial air pollution.*

Nitric oxide (NO), which also contributes to the natural acidity of rainwater, is formed during lightning storms by the reaction of nitrogen and oxygen, two common atmospheric gases.

In the air, NO is oxidized to nitrogen dioxide (NO_2), which in turn reacts with water to give *nitric acid* (HNO_3). This precipitation has a damaging effect on plants, animals, and infrastructure.

Ground-level ozone (O_3) is formed when VOCs chemically react with nitrous oxides NO_x in the presence of sunlight. Sources of these pollutants are gasoline vapors, emissions from industrial activities, motor vehicle exhaust, chemical solvents, and electric utilities. The effects of these pollutants on humans are heart problems and respiratory symptoms, including lung diseases, asthma, shortness of breath, and coughing. They also have adverse effects on sensitive vegetation and fragile ecosystems.

Measurement units

Air quality measurements are typically reported in micrograms per cubic meter ($\mu g/m^3$) and parts per million (ppm) or parts per billion (ppb). Particulate matter sizes are expressed in micron or micrometer.

Smog

Smog was initially used to describe the combination of human-made smoke and fog in 20th century London. In the contemporary vernacular, *smog* refers to the pollution that results from the excessive burning of coal or an excess of motor vehicle exhaust mixed with fog. Smog is also formed when vehicle exhaust reacts with ultraviolet light in the atmosphere, forming secondary pollutants that combine with primary pollutants to form photochemical smog.

Figure 6.3. *Industrial pollution and smog.*

Smog is typically produced through a complex set of photochemical reactions between *volatile organic compounds* (VOC) and nitrogen oxides in the presence of sunlight. This reaction forms ground-level ozone and is an issue in industrialized cities.

The sources of these pollutants include automobile exhausts, industrial emissions, hair spray, solvents, coal fires, and paint. The main pollutants in urban areas originate from vehicle emissions. This is very toxic to humans and causes coughing, difficulty breathing, asthma, colds, lung infections, choking, and eye irritation. The effects on plant life are the restriction of plant growth and damage to forests and crops.

Acid deposition – causes and effects

Acid deposition is the release of acidic or acid-forming pollutants in the atmosphere on the Earth's surface. The deposition occurs as both wet and dry deposits. Wet deposits are precipitation in the form of rain, snow, sleet, mist, and in some cases, fog. Dry deposition includes dust or smoke and falls to the ground as dry particles.

The main pollutants responsible for acid deposition are sulfur dioxide and nitrogen oxides. These are emitted from vehicle exhaust, electric utility plants, and power plants. These primary pollutants, once emitted, react with water and oxygen to produce sulfuric acid and nitric acid. These compounds dissolve very easily in water and can be carried long distances by the wind, remaining airborne for long periods before depositing on the Earth's surface.

Acid deposition lowers the pH of the soil. This results in the leaching of nutrients, which in turn affects the health of plants. This occurred in New York in the Adirondacks, where widespread cations were leached from the forest floor, resulting in increased levels of sulfur, nitrogen, and calcium deficiency in the soil. This led to a decrease in the plant and animal diversity of the forest. Useful microorganisms are also killed due to acid deposition, which ultimately leads to infertile soil.

Acid deposition can fall directly on bodies of water or be introduced to a body of water through run-off from terrestrial environments. Acid deposition within a body of water lowers the pH balance of the aquatic ecosystems, which is hostile to aquatic life. The result is that aquatic animals and plant species die, as was the case with the New Jersey Pine Barrens.

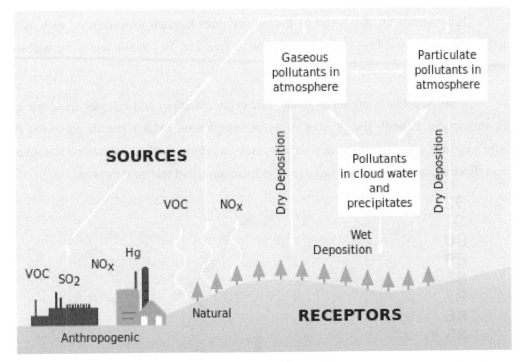

Figure 6.4. *Acid deposition.*

The effect of acid deposition on forests is that trees become susceptible to diseases and their growth becomes stunted. Severe forest damages have also been reported due to acid depositions in areas of Eastern Europe, specifically in Germany, Poland, and Switzerland.

Acid deposition also affects property. Buildings constructed with limestones, such as historic buildings in Europe and older settlements in the Northeast of the U.S., are eroded by acid rain. Cars, bridges, and airplanes are also susceptible to corrosion and weakening of the structure.

Public health is affected by acid deposition as sulfur dioxide, nitrogen oxide gases, and particulate matter reduces visibility, causing accidents and leading to injuries or deaths.

Heat islands and temperature inversions

Heat islands are created when hot air layers or temperature "domes" form over an urban or industrial area. This dome is typically warmer by 5-7 °C than the surrounding air and forms a trap for air pollutants. Heat islands are most recognizable in the early mornings and late nights but disappear during the day because of the overall temperature increase of the atmosphere.

Heat islands are the result of human activities through urbanization, such as the modification of natural land surfaces to asphalt or concrete. Vegetation and soil can absorb radiation.

Also, vegetation produces shade, intercepts radiation, and releases moisture into the atmosphere through the process of evapotranspiration, which are all processes that cause a cooling effect. Humanmade surfaces lack this ability as they are water-resistant and non-reflective, causing these surfaces to absorb radiation and release it as heat.

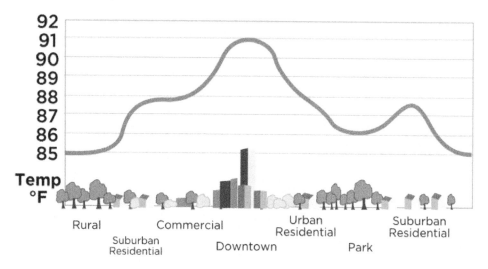

Figure 6.5. *Urban heat island profile.*

The heat generated from energy usages, such as from air conditioning systems, industrial processes, and refrigeration, also contributes to the formation of temperature domes. Buildings can obstruct the normal flow of air and lead to an increase in temperature.

Temperature inversion refers to an atmospheric condition that occurs when cold air gets trapped under a layer of warm air. This is related to an inversion because the normal decrease of temperature with height is "inverted" so that there is an increase in temperature with height. This inversion acts as a seal that prevents connective overturning of air currents. Air pollutants, which normally would have dispersed over a wide area, become trapped below the inversion. This phenomenon has the potential to affect weather patterns.

Indoor air pollution

Indoor air pollution is when pollutants from gases and particles contaminate the air indoors. Indoor air is far more concentrated with pollutants than outdoor air. Indoor air pollution has a greater impact on developing countries than in developed countries. This is largely because, in developing countries, fuels (e.g., wood and charcoal) are burned inside homes for cooking and heating.

The poor ventilation in most of these homes prevents pollutants from escaping, causing the residents to breathe in carbon monoxide and other dangerous contaminants.

Indoor fuel burning leads to human health problems such as pneumonia, bronchitis, cancer, heart disease, and asthma. Dangerous indoor air pollutants in developed countries are tobacco smoke, radon, and carbon monoxide. Second-hand smoke has the potential to cause lung cancer, emphysema, asthma, and heart disease.

Radon is a carcinogen gas that is exuded from the Earth in certain locations and trapped inside houses. Volatile organic compounds are also common indoor pollutants. VOCs are released by many household items and everyday products, such as cologne, paint, plastic, solvents, pesticides, carpets, and furniture.

Figure 6.6. *Chamber for measuring volatile organic compounds emitted from furnishings.*

When VOC are released indoors, they are more concentrated than when they are released into an open space. Biological sources such as dust mites, pet dander, mold, mildew, and airborne bacteria are also indoor pollutants that can irritate the human respiratory system. Materials used in building such as asbestos, lead, and formaldehyde are indoor pollutants. Indoor pollution can be reduced by providing adequate ventilation in buildings as well as ridding the space of the source of the pollutants.

Remediation and reduction strategies

Air pollution remediation is the removal of air pollutants. Remediation is usually based on various regulatory requirements and assessment of human health and environmental risks where no legislated standards exist or where standards are merely advisory. Commercial operations can contribute to the reduction of air pollution by using more sustainable forms of transportation such as electric, gas hybrid, clean fuel, or emissions-free vehicles. Restaurants can use natural gas or electricity for preparing food instead of using open pit grills. Air pollution from the residential sector can be reduced by eliminating indoor fireplaces.

Environmentally friendly solvents and household cleaners can be used instead of environmentally harmful ones. Sustainable transportation options include walking, carpooling, riding a bike or taking public transportation. Efforts to reduce pollution from mobile sources include manufacturing hybrid vehicles, making cleaner fuels or using electric-powered vehicles.

Air pollution is largely the result of burning fossil fuels such as coal and oil. This contribution can be significantly reduced by using renewable energy sources, including wind power, solar power, and hydropower.

Control devices can be used to reduce air pollution by destroying pollutants or removing them from exhaust streams before they are released into the atmosphere from equipment.

Clean Air Act and other relevant laws

There are two types of air quality standards. The first class of standards, such as the U.S. National Ambient Air Quality Standards, sets maximum threshold limits for atmospheric concentrations of specific pollutants. The second type of air quality standards, such as the North American Air Quality Index, are scales with various thresholds used to communicate the relative risk of outdoor activity to the public; the scales may or may not distinguish between different pollutants.

Air quality laws regulate the emission of air pollutants into the atmosphere and are usually specifically designed to protect human health by limiting or eliminating airborne pollutants emitted. Some initiatives are aimed at addressing ecological problems, such as reducing the emissions of chemicals that affect the ozone layer and programs to address acid rain or climate change issues. Regulatory efforts include identifying and categorizing air pollutants, setting limits on acceptable standards, and suggesting mitigation strategies and technologies that should be implemented.

The *Clean Air Act* came into effect in 1963, and it is a federal law designed to govern and regulate the reduction of air pollution. This law is administered by the U.S. Environmental Protection Agency, in coordination with state, local, and tribal (Native American) governments. The EPA sets standards for levels of pollutants in the air, such as ozone, particulates, and nitrogen oxides in smog.

Each state has the responsibility to ensure that these standards of the Clean Air Act are met. Efforts to keep air quality to standards include strategies to control emissions, such as changes in the composition of gasoline, the use of alternative fuels such as natural gas and electricity and banning charcoal barbecues and wood-burning stoves or fireplaces where pollution levels are high.

Also, carpooling is encouraged, and efforts are aimed at managing traffic in congested areas, improving public transportation systems, encouraging employers to contribute to employee mass transit costs, and implementing "smog fees" on cars in proportion to the distance traveled and vehicle emissions produced.

Figure 6.7. *President Lyndon B. Johnson signed the Clean Air Act in 1967.*

The *Air Quality Act* of 1967 was an amendment to the Clean Air Act of 1963. The Air Quality Act enabled the federal government to increase its activities in the investigation and enforcement of interstate air pollution transport and, for the first time, to perform far-reaching ambient monitoring studies and stationary source inspections.

The 1967 act also authorized expanded studies of air pollutant emission inventories, ambient monitoring techniques, and control techniques.

The *Air Pollution Control Act*, which came into effect in 1955, was the first U.S. federal legislation that related to air pollution. This act also provided funds for federal government research on air pollution.

The *Air Pollution Control Act* identified air pollution as a national issue and announced that research and more ways to work toward improvement were needed to tackle the issue at hand. This act was geared at increasing the awareness of the environmental hazard of air pollution.

Noise pollution

Noise pollution refers to excessive noise or unpleasant sounds that disrupt the natural balance.

Sources of noise pollution

Outdoor sources of noise are mainly caused by machines and transportation systems, motor vehicles, aircraft, and trains. Urbanization is a contributor to noise pollution as industrial and residential buildings are in close proximity. Indoor noise pollution is caused by machines and equipment, especially in factories.

Effects of noise pollution

The most common health effects associated with high levels or long periods of noise exposure are hearing loss, but other adverse effects may include psychological problems, tinnitus, stress-related illnesses, high blood pressure, sleep disruption, cardiovascular diseases, and a decrease in productivity.

Control measures

Steps can be taken to prevent the harmful effects of noise pollution, such as wearing ear protection (like earplugs or earmuffs), when areas with loud noise levels cannot be avoided, as in occupational settings.

Size and alteration of roadway surface texture, traffic controls to reduce vehicle speed, and noise barriers can be implemented to reduce noise. Residents who live close to airports would benefit from aircraft with quieter engines and strategically chosen flight paths based on the time of the day.

Water pollution

Water pollution is the contamination of natural bodies of water such as lakes, rivers, oceans, aquifers, and groundwater by particles or chemicals from human activities.

Types of water pollution

Water can become polluted in several ways:

Nutrient pollution occurs when bodies of water become contaminated by wastewater, fertilizers, or sewage that contains high levels of nutrients. This contamination causes excessive weed and algae growth in the water, making the water unpotable. High levels of algae expend all the oxygen in bodies of water, starving other aquatic organisms.

Figure 6.8. *Nutrient pollution: runoff of fertilizer during heavy rain.*

Surface water pollution is the contamination of natural bodies of surface water, such as rivers, oceans, lakes, and lagoons.

Oxygen depletion of water occurs when aerobic and anaerobic organisms feed on biodegradable material. This causes biodegradable matter to build up in the water, encouraging microorganism growth and expending the available oxygen in the water.

When oxygen is depleted, the aerobic organisms die, and anaerobic organisms begin to reproduce, producing harmful toxins such as ammonia and sulfides.

Groundwater pollution occurs when chemical contaminants, such as pesticides, are leached from soils and washed underground and into aquifers by rainwater. This can affect drinking water supplies that are accessed through ground wells.

Microbiological water pollution is a natural form of water pollution caused by the presence of numerous varieties of microorganisms such as viruses, bacteria, and protozoa. This microorganism population can kill aquatic species of animals and plants and cause serious illness, including the transmission of water-borne diseases (i.e., cholera) to the humans who drink this water.

Suspended matter is a form of water pollutant characterized by chemicals and particles suspended in bodies of water which are insoluble or have a low solubility. Some suspended particulate matter may eventually settle under the water body forming a layer of silt, endangering the aquatic life that lives on the floor of these bodies of water.

Chemical water pollution occurs when chemical runoff from farms and factories enter bodies of water, polluting them. These chemicals are poisonous to marine life and, if consumed by birds or humans, can infect the consumer.

Oil spills are a form of water pollution resulting from the spilling of oil into a body of water. These spills can spread across a large area and affect numerous forms of wildlife. They can kill fish and cause birds to lose their ability to fly as oil sticks to their feathers.

Sources, causes, and effects of water pollution

The two main sources of water pollution are point and non-point sources. *Point sources* are pollutants that belong to a single source, such as emissions from factories. *Non-point sources* are pollutants emitted from multiple sources, for example, water that has traveled through several regions and picked up different contaminants.

Industrial waste, such as lead, mercury, sulfur, asbestos, nitrates, and other harmful chemicals, is produced as waste from industrial activities. Many industries do not practice proper waste management and drain their waste into water systems. This has the potential to cause eutrophication, damaging aquatic life.

Sewage and wastewater from homes are treated and released in the sea. Sewage water contains numerous pathogens that are harmful to human health and the environment.

Mining activities include the extraction of minerals, some of which contain harmful chemicals that can be released and get into the groundwater supply. This waste includes sulfides, which are toxic and may cause health problems.

Marine dumping occurs when garbage is disposed of in the sea. Some materials take hundreds of years to decompose. This not only pollutes the water but harms marine plants and animals as well.

Figure 6.9. *Marine debris on a Hawaii beach.*

Oil spills can occur when a ship or train carrying large oil containers crashes or a pipeline carrying oil ruptures. Any of these occurrences are cause for major environmental concern due to the threat of oil spilling into the ocean or the water table. An oil spill may affect marine wildlife such as fish, birds, and sea otters. The severity of the damage is dependent on the quantity and toxicity of the oil spill and the size of the body of water in which the spill occurred.

Chemical fertilizers and pesticides are utilized by farmers to protect crops from insects, weeds, and microorganisms such as bacteria. These chemicals may contact bodies of water due to rainwater runoff, posing a serious threat to aquatic animals.

Radioactive waste, produced from radioactive materials (e.g., uranium used to produce nuclear energy), is a highly toxic chemical. Nuclear waste needs to be disposed of carefully to prevent nuclear accidents; nuclear waste can have detrimental environmental implications.

Landfills are massive, layered piles of garbage. Rainwater that percolates through a landfill may end up carrying pollutants that eventually meet up with groundwater if the landfill is not properly lined. This will pollute the groundwater supply, which may affect the public drinking water supply.

Figure 6.10. *Industrial scrap yard as a source of groundwater pollution if not lined properly*

Animal waste can be washed into rivers through rainfall and then mixed with other harmful chemicals. This has the potential to cause many different types of water-borne diseases such as cholera, diarrhea, jaundice, dysentery, and typhoid.

Water pollution has a devastating effect on ecosystems, human health, and even tourism and recreational activities. This pollution affects sea creatures and accumulates in their bodies, the negative effects of which are then passed on to humans when they eat the fish. Water pollution harms the environment on which humans are dependent for food, oxygen, and water.

Cultural eutrophication

Cultural eutrophication occurs when anthropogenic activities accelerate the process of eutrophication by causing excess nutrients, such as phosphates and nitrates from sewage, detergents, and fertilizers to enter the ecosystem. These added nutrients cause an overgrowth of algae, resulting in increased competition among other marine organisms for sunlight and oxygen. This decreases the dissolved oxygen in the water, which eventually kills other aquatic plants and animals. This situation can cause the water to become unpotable.

Groundwater pollution

Groundwater pollution occurs when pollutants are released to the ground and seep into the water table. The pollutant creates a contaminant plume within an aquifer. Water currents quickly disperse the pollutant over a wide area. This pollutant can seep into springs and wells, contaminating the potable water supply.

Groundwater pollution can occur from improperly maintained sanitation systems, landfills, effluent from wastewater treatment plants, leaking sewers, gas stations, air pollutants, or the overuse of fertilizers in agriculture.

Contamination can also occur from naturally occurring elements in soil, such as arsenic or fluoride. Drinking polluted groundwater may harm public health, plants, and animals through poisoning or spreading disease.

Figure 6.11. *Groundwater pollution: a pit latrine in Lusaka, Zambia pollutes the nearby well.*

More than half of the United States population depends on groundwater for drinking water. Groundwater is also largely used for irrigation purposes. Drinking contaminated groundwater can have adverse health effects.

Diseases such as hepatitis and dysentery are caused by the consumption of water contaminated with toxins that have been leached from septic tank waste.

Wildlife can also be harmed by contaminated groundwater. Other long-term effects include many cancers, which are linked to both long and short-term exposure to polluted water.

Maintaining water quality

Water is essential to human life and environmental health. This valuable natural resource includes marine, estuarine, freshwater, and groundwater environments. Water has two attributes that are closely linked: quantity and quality.

Water quality refers to its physical, chemical, biological, and aesthetic characteristics. A healthy environment is one that supports water quality, which in turn supports a diverse community of organisms and is safe for human consumption. Water quality is important because it assists in maintaining public health and ecosystems and is important in industries such as farming, fishing, mining, and tourism.

Water quality is affected by public uses such as agricultural, urban, industrial, and recreational applications. Modification of the natural path of streams using dams and dikes may also affect water quality. The weather can also have an impact on water quality, especially in dry seasons.

The quality of water is measured by the level of contaminants and its characteristics. The chemistry of water can be affected by pH, dissolved oxygen, biological oxygen demand, minerals, and organic and inorganic compounds.

The biological characteristics of water include the level of microorganisms and algae. The physical characteristics of water include the temperature, turbidity and clarity, color, salinity, suspended solids, and dissolved solids. The aesthetic qualities used to measure water quality have to do with its odors, taints, color, and floating matter.

The radioactive component of water includes alpha, beta, and gamma radiation emitters. Closely monitoring these indicators is necessary to determine if the water is suitable for consumption.

Water quality can be improved by providing information to the public on the quality of water through programs and environment reports. The public should also be educated on the importance of maintaining water quality. Pollution reduction programs can be developed, as well as industrial and water quality management strategies.

Water purification

Water purification is the process by which chemical, biological, and suspended solid contaminants are removed from the water. This is done to render water suitable for a specific use, such as human, industrial, medical, pharmacological, and chemical applications.

This process removes bacteria, algae, viruses, fungi, lead, copper, and suspended particles. A small amount of disinfectant is usually left in the water at the end of the treatment process to reduce the risk of re-contamination once in the distribution system.

Boiling water is a simple procedure that can be used to treat water. Other physical methods used to purify water include filtration, sedimentation, and distillation. Biological processes used for purification include slow sand filters or biologically active carbon. Chemical processes include flocculation, chlorination, and the use of electromagnetic radiation, such as ultraviolet light.

Sewage treatment and septic systems

Wastewater from flushing the toilet, bathing, and washing sinks go down the drain and into pipes, which join up with larger sewer pipes that lead to a treatment center.

There are four stages of sewage treatment, which include: screening, primary treatment, secondary treatment, and final treatment.

Figure 6.12. *A sewage treatment plant in Germany.*

Screening is the first stage of the wastewater treatment process. This stage includes the removal of large objects such as plastics, rags, diapers, sanitary items, cotton buds, and face wipes that have the potential to block or damage equipment.

The *primary treatment stage* involves the separation of human waste from the wastewater. Wastewater is moved into large settlement tanks where the solids sink to the bottom of the tank to form sludge. Large scrapers at the bottom of these circular tanks scrape the floor and push the sludge towards the center where it is pumped away for further treatment. The rest of the water is then moved to secondary treatment.

The third stage is where the *secondary treatment stage* takes place. The water is placed into large rectangular tanks called *aeration lanes*. Air is pumped into the water to encourage bacteria to break down the tiny bits of sludge that escaped the sludge scraping process.

In the *final treatment stage*, the wastewater is passed through a settlement tank. More sludge forms at the bottom of the tank through the settling of the bacterial action. The sludge is again scraped and collected for treatment. The water is then allowed to flow over a wall for further filtration through a bed of sand, which removes additional particles. The filtered water is then released into the river.

Clean Water Act and other relevant laws

The *Clean Water Act* (CWA) is the federal law responsible for regulating water pollution in the United States. The purpose of this act is to maintain and restore the chemical, physical, and biological standards of the nation's water. The CWA was passed in 1948 under the name of the *Federal Water Pollution Control Act*. The act gained significant recognition in 1972, at which time it was amended. The CWA is administered by the EPA in coordination with state governments.

At both the federal and state levels, this law allows for and encourages the implementation of strategies and programs designed to control pollution, such as setting wastewater standards for the American industries. Water quality standards have also been set for all contaminants in surface waters. The CWA has made it illegal to discharge pollutants from any point source into navigable waters without a permit. EPA has a *National Pollutant Discharge Elimination System* (NPDES) permit program that controls discharges.

The *Safe Drinking Water Act* (SDWA) is the main federal law that assures the quality of Americans' drinking water. Under the SDWA, standards are set by the EPA for the expected quality of drinking water and manages the states, localities, and water suppliers who implement those standards. This act was originally passed in 1974 by Congress to protect the public's health through the regulation of the nation's public drinking water supply.

The SDWA was further amended in 1986 and 1996, and it now requires many actions to protect drinking water and its sources, including rivers, lakes, reservoirs, springs, and groundwater wells. However, the SDWA has no regulatory authority over the standards and use of private wells that service the needs of about 15% of the population.

Under the requirements of the SDWA, the EPA sets national health-based standards for drinking water to protect against naturally occurring and human-made contaminants that may be found in potable water.

The threats to drinking water can include improperly disposed of chemicals, animal waste, pesticides, human waste; waste injected deep underground and naturally occurring substances. This act is important to maintaining human health, as drinking improperly treated or disinfected water, or water which travels through an improperly maintained distribution system, has the potential to pose serious health risks.

Solid waste

Solid waste is a solid or semisolid, non-soluble material that is discarded or abandoned. Solid waste can include agricultural refuse, demolition waste, industrial waste, mining residues, municipal garbage, sewage sludge, and food waste.

Types

There are three types of solid waste: municipal or household waste, industrial or hazardous waste, and biomedical or hospital waste.

Municipal solid waste is mainly comprised of garbage generated from residential and commercial processes. Increasing urbanization and changes in lifestyle have resulted in a similar change in the amount and composition of municipal solid waste.

The packaging of products has also seen a change as aluminum foil, plastics, cans, and other non-biodegradable materials are commonly used. The municipal waste consists of household waste, construction and demolition debris, sanitation residue, and waste from streets.

Hazardous waste is toxic waste, such as industrial and hospital waste. Household waste such as old batteries, paint, expired medications, and shoe polish are also classified as hazardous waste. These wastes have the potential to be highly toxic to humans, animals, and plants due to their corrosive, flammable, explosive, and reactive properties.

Biomedical waste can also be considered hazardous when contaminated by chemicals such as formaldehyde and phenols. These chemicals are readily found in disinfectants, the mercury used in thermometers, and equipment used to measure blood pressure.

The main sources of industrial waste are the chemical, metal, and mineral processing industries. Industries generate toxic waste from various processes involving metals, chemicals, paper, pesticides, dyes, refining, and rubber goods.

Direct exposure to chemicals in hazardous waste such as mercury and cyanide can be fatal. Nuclear plants generate radioactive wastes, thermal power plants produce fly ash, and chemical industries generate large quantities of hazardous and toxic material.

Figure 6.13. *Hazardous waste collection site.*

Biomedical waste is generated during the diagnosis, treatment or immunization of humans or animals, in research activities in the medical field and the production or testing of biologicals. This type of waste is highly infectious and can cause a serious threat to human health if not properly managed in a scientific and discriminate manner.

It has been estimated that one kilogram of every four kilograms of waste generated in a hospital or other medical facility is infected. Biomedical waste includes sharps, soiled waste, disposables, anatomical waste, cultures, discarded medicines, chemical wastes, etc. These are in the form of disposable syringes, swabs, bandages, body fluids, and human excrement.

Disposal of solid waste

Solid waste is disposed of using: landfills, incineration, and composting.

A *sanitary landfill* is the most economical method of garbage disposal, and it is created by alternating layers of thin, compacted, and solid waste covered with clay or plastic foam. The bottom of the landfill is sealed with an impermeable material such as clay, plastic foam, and sand. This impermeable layer protects the groundwater from seepage from the landfill.

When the landfill is full, it is covered with clay, sand, gravel, and topsoil to prevent the seepage of water. Areas close to the site are drilled and monitored to prevent groundwater contamination. In two to three years, the solid waste volume shrinks by 25-30%, and the land is returned to other uses for the construction of roads, small buildings, and parks.

The advantages of using this method to dispose of solid waste are that it is an easy and economical method, and the wastes need not be grouped based on similarity. Also, this method makes use of low-lying land and swamplands which can be reclaimed and used for other purposes.

Natural waste, which is biodegradable, is also returned to the soil and provides nutrients. The disadvantage of this method is that it demands the use of a large area of land, which can be costly or unavailable in densely urbanized areas. This method also requires constant monitoring to prevent foul odors, groundwater seepage and disease-carrying vectors such as mosquitoes, roaches, rats, and flies.

Landfills also have the potential to cause fires due to the formation of methane gas from decaying organic waste in wet weather conditions.

Solid waste disposal using incineration involves the burning of solid, liquid, and gaseous waste until it is reduced to ashes. Incinerators are furnaces that are well insulated to prevent the escape of excess heat and pollutants. This process reduces the volume of solid waste to approximately 20-30% of its original volume.

This is a practical method of disposing of some types of hazardous waste materials, but there are concerns about the effects this method may have on air quality.

Garbage is separated into non-combustible substances and non-combustible matter. Non-combustible substances can be rubbish, garbage, and dead organisms. Non-combustible matter, for example, glass, porcelain, and metals are separated and may be recycled or reused. The residue or ashes from incineration requires further disposal by sanitary landfill or some other means.

The advantages of using this method include the fact that an incinerator requires a small space and the residue be much smaller than its original volume. An incinerator plant with a capacity to dispose of 3,000 tons of garbage a day can generate three megawatts of power.

The heat generated during combustion is used in the form of steam power for the generation of electricity through turbines. This method of waste disposal is not without its limitations, which include high capital and operating cost and the requirement of skilled personnel to operate the incineration process. Also, smoke, dust, and ashes produced during incineration require management and further disposal to prevent air pollution.

Composting is popular in many urban areas where there is a shortage of landscape for landfills. In this method, organic waste is naturally broken down into fertilizer by microorganisms, especially bacteria and fungi.

Biodegradable waste is dumped in underground trenches in 1.5-meter layers covered with earth and left for decomposition. This decomposition forms humus, which has a high level of nitrogen and can be used in agriculture for plant growth to replace chemical fertilizers.

In *vermicomposting*, worms are added to the compost pile. These organisms assist with the breaking down of waste, and the added excreta of the worms make the compost very rich in nutrients. The benefits of composting include the recycling of nutrients that are returned to the soil in a clean, cheap, and environmentally safe method.

A disadvantage of this method is that non-biodegradable waste must be disposed of by other means.

Reduction

Reducing the amount of waste produced has a positive impact on the environment. This results in less garbage going to landfills.

Reduction in waste can be achieved by employing the following practices:

1) Using a reusable bag(s) when shopping;

2) Buying products with less packaging or environmentally friendly packaging;

3) Starting a backyard compost pile using organic kitchen waste;

4) Reusing items that are not biodegradable;

5) Recycling as much as possible;

6) Purchasing recycled products;

7) Borrowing/renting rarely used items instead of purchasing them;

8) Avoid traveling by car alone or over short distances, using more sustainable forms of transportation where possible (e.g., public transportation, walking, riding a bicycle);

9) Using a computer instead of buying hard copy newspapers and magazines, printing only what's necessary;

10) Turning off lights and appliances that are not being used; and

11) Turning off the faucet while brushing teeth and lathering in the shower.

Notes

Impacts on the Environment and Human Health

The increase in population has led to an increase in the consumption of food and other items essential to everyday life. Generated waste is discarded into municipal waste collection receptacles from whence it is collected by the area municipality to be further disposed of into landfills and dumps. These municipalities sometimes suffer from a lack of resources or an inefficient infrastructure, which can lead to failure in waste collection.

Waste that is not properly managed, especially excreta and other liquid and solid waste from households and the community, can have serious health and environmental effect. Left unattended, standing waste may attract insects and vermin which carry disease and contribute to poor living conditions.

Hazards to human health

The main population at risk from the build-up of waste is people living in areas where there is no proper waste disposal method. Populations living near waste dumps and those whose water supply has become contaminated due to either waste dumping or leakage from landfill sites are also at risk. Uncollected solid waste also increases the risk of injury and infection.

Organic waste poses a serious threat because, as this waste decomposes, it creates conditions that encourage the survival and growth of microbial pathogens. Waste disposal workers are most vulnerable to various types of infectious and chronic diseases from direct handling of solid waste.

The disposal of industrial hazardous waste alongside municipal waste can expose people to hazardous chemical and radioactive substances. Chemicals such as cyanide, mercury, and polychlorinated biphenyls are highly toxic, and exposure can lead to diseases like cancer and death. Uncollected solid waste can also obstruct stormwater runoff, resulting in the forming of stagnant bodies of water that become breeding grounds for disease.

The dumping of waste near a body of water or groundwater can cause contamination, which affects the human water supply. Also, direct dumping of untreated waste in rivers, seas, and lakes results in the accumulation of toxic substances in the food chain through the plants and animals that feed on it.

Improperly operated incineration plants cause air pollution, and improperly managed and designed landfills attract all types of disease-carrying vectors. These sites should be located a safe distance from all human settlements. Landfill sites should be well-lined and solidly walled to prevent leakage into nearby groundwater sources.

Figure 6.14. *Mercury waste from the Brunswick Pulp and Paper Company has destroyed the marshland in front of the plant.*

Disposal of hospital and other medical waste requires special attention. The waste generated from medical facilities such as discarded syringe needles, bandages, swabs, adhesive bandages, and other types of infectious waste contains toxic chemicals and metals.

Contact with these infectious items may result in major health hazards (such as Hepatitis B and C) through wounds caused by discarded syringes.

Proper methods of waste disposal will ensure that this practice does not affect the environment or cause health hazards to people living near disposal sites.

Air pollutants cause asthma, a disease that affects the lungs. It is characterized by repeated episodes of wheezing, breathlessness, chest tightness, and nighttime or early morning coughing. This disease can be controlled by medication and by avoiding the triggers that cause it.

Pollution has also been linked to cancer clusters, which are larger-than-expected numbers of cancer cases occurring within a group of people in a geographic area over some time.

One of the most infamous of such events occurred in the 1960s with the case of mesothelioma, a rare cancer of the lining of the chest and abdomen. Studies traced the development of this cancer to exposure to asbestos.

Carbon monoxide is produced through the combustion of fossil fuel. This gas can cause sudden illness and death. This has been managed in the United States by raising awareness about carbon monoxide poisoning and monitoring the illnesses and deaths resulting from this gas.

Environmental risk analysis

As water encounters decomposing solid waste, it dissolves with the soluble inorganic and organic wastes, producing a polluted liquid known as leachate. Solid waste affects water quality through the release of leachate from landfills into water sources.

Leachate becomes more concentrated as it seeps into deeper layers of the landfill; this contributes to the light brown/black color of leachate and its pungent odor. It has a high polluting potential impact due to the high concentrations of organic contaminants and ammonia nitrogen.

Leachate discharged into a body of water will have an acute and continuing impact. If toxic metals are present, this can lead to chronic toxin accumulation in organisms that depend on it, such as fish, and may consequently affect the humans who feed on these organisms.

Figure 6.15. *Workers test a leachate collection system at the Savannah River Site.*

Environmental risk due to water pollution is an ongoing problem. Municipal wastewater systems are one of the largest sources of pollution to surface waters. Other major sources include residential and industrial discharges and agricultural runoff. Chemicals from pesticides may also contaminate bodies of water. Air pollution and land pollution also affect water quality.

Acute and chronic effects

When chemicals contaminate a body of water, this can affect the surrounding wildlife, watershed, and residents. For instance, if chemicals find their way into a freshwater source that supplies people or animals, it may no longer be safe for use or consumption. Toxic release from industrial plants and agricultural runoff into the environment have a short-term effect on water supplies.

Long-term effects of water pollution include contamination of bodies of water by chemicals from fertilizers and sewage. Nitrate and phosphate from fertilizer provide food for algae resulting in algae blooming. When the excess algae die and decay, they use the existing, dissolved oxygen and degrade the overall quality of the water. This oxygen deprivation kills other aquatic life.

Sulfur and nitrogen oxide emissions from industrial plants can cause acid rain, which affects both aquatic and terrestrial plants and animals and leeches the soil of essential nutrients.

Long-term exposure to chemical pollutants kills native species within an ecosystem, resulting in a loss of biodiversity and causing the area to become vulnerable to invasive and undesirable species.

Dose-response relationships

Dose-response relationship (also referred to as exposure-response relationship) describes the change experienced by an organism resulting from different levels of exposure to a stressor over a specified period.

Pollutants mainly enter the human body through breathing. Pollutants also find their way into the body through ingestion (e.g., children eating soil contaminated with lead) or are absorbed through the skin. Once a pollutant such as asbestos enters the body, it can stay in the lungs, be exhaled, or be absorbed into the bloodstream through the lungs, stomach, or skin. Through the bloodstream, it gets transported to different parts of the body.

Pollutants can undergo chemical changes as they are transported throughout the body, especially in passing through the liver. This mutation can cause the pollutant to decrease or increase in toxicity level. Pollutants can also be excreted from the body through urination, bowel movements, breast milk, or perspiration. They can also be stored in bone, fat, organs, or hair.

Toxic air pollutants affect normal body functions by changing chemical reactions within cells. These changes can destroy cells, damage cell function or alter cell activity and result in impaired organs, congenital disabilities when the cells of an unborn child are damaged or cancer that can develop when cells begin to grow at an uncontrolled rate.

The dose-response relationship for a specific pollutant refers to the relationship between exposure level and the corresponding health effect.

The EPA assumes that no exposures have "zero risks;" even significantly low exposure to a carcinogenic pollutant can increase the risk of cancer even by a small degree. The EPA also assumes that the relationship between dose and response is a straight line; for each unit of increase in exposure, there is a corresponding increase in cancer response.

The EPA has also determined an exposure level that exists at or below the minimum health effect level and asserts that this low exposure will not provide any real harm to the body's natural protective mechanisms, which can repair any damage caused by a pollutant.

The dose-response relationship varies with the type of pollutant, individual sensitivity, and type of health effect. Studies reveal that long-term exposure to low concentrations of particulate matter from combustion is associated with chronic health effects such as bronchitis, reduced lung function, shortened lifespan, lung cancer, and increased rates of respiratory symptoms.

Figure 6.16. *The United States Environmental Protection Agency logo.*

Air pollutants

The most significant air pollutants from the human health standpoint are particles, acidic gases, aerosols, metals, and organic compounds resulting from incineration. Exposure to these particles has acute health effects, such as increased overall mortality and emergency hospital admissions.

Acidic gases such as SO_2 have adverse effects on lung functions, especially in asthmatics. Environmental exposure to particulate matter is associated with increased cardiovascular and respiratory mortality and morbidity.

Metals from incinerator emissions such as lead, cadmium, mercury, chromium, arsenic, and beryllium have both carcinogenic and non-carcinogenic health effects. The organic compounds derived from incineration, such as *dioxins* and *polychlorinated biphenyls* (PCBs), have the potential to accumulate in the body. High levels of dioxin exposure found in workplaces and after accidents have caused chloracne and an increase in cardiovascular disease.

Smoking and other risks

Cigarettes contain about 600 ingredients, and when they burn, they generate more than 7,000 chemicals, including nicotine, tar, carbon monoxide, formaldehyde, ammonia, hydrogen cyanide, arsenic, and dichlorodiphenyltrichloroethane (DDT). Many of those chemicals are poisonous, and at least 69 of them are carcinogenic. Many of the same ingredients in cigarettes are found in cigars and the tobacco smoked in pipes and hookahs.

The National Cancer Institute reported that cigars have a higher level of carcinogens, toxins, and tar than cigarettes. According to the Center for Disease Control and Prevention, the mortality rate for smokers in the United States is three times that of people who never smoked. Smoking is one of the leading causes of preventable death. Smoking has been found to harm nearly every bodily organ and organ system in the body.

Nicotine reaches the brain seconds after it is inhaled. It is a central nervous system stimulant with high habit-forming potential. Smoking increases the risk of macular degeneration, cataracts, and poor eyesight. It can also weaken the senses, including taste and smell.

Inhaling smoke over time will damage the lungs, which will lose their ability to filter harmful chemicals. Coughing is not an adequate response to clear out the toxins, causing these toxins to become trapped in the lungs. This results in smokers having a higher risk of respiratory infections, colds, and flu.

Over time, smokers are at an increased risk of developing forms of chronic obstructive pulmonary disease, such as emphysema and chronic bronchitis. Emphysema is characterized by the air sacs in one's lungs being destroyed by harmful chemicals. In chronic bronchitis, the lining of the tubes of the lungs becomes inflamed. Long-term smokers are at an increased risk of lung cancer.

Smoking damages the entire cardiovascular system. Nicotine results in the tightening of blood vessels, which restricts blood flow. Smoking additionally lowers the level of good cholesterol in the blood and raises blood pressure, which can result in stretching of the arteries and an accumulation of bad cholesterol, leading to atherosclerosis.

Smoking also increases the risk of blood clots, which along with weakened blood vessels in the brain, increase the risk of stroke. A long-term effect of smoking is a greater risk for leukemia. Inhaling second-hand smoke increases the risk of stroke, heart attack, and coronary heart disease even in non-smokers.

More obvious consequences of smoking are the effects it has on the skin including discoloration, wrinkles, and premature aging. Substances in tobacco smoke change skin structure. Fingernails and the skin on fingers may also develop yellow staining from holding cigarettes. Smokers usually develop yellow or brown stains on their teeth.

Smokers are at an increased risk of oral health problems. Tobacco has the potential to cause gum disease, gingivitis, or infections such as periodontitis. These problems can develop into other issues such as tooth decay, tooth loss, and bad breath. Smoking also increases the risk of developing cancer of the mouth, throat, larynx, and esophagus. Smoking also increases the risk of kidney and pancreatic cancers.

Smoking increases the chance of developing insulin resistance, which makes it more likely for smokers to develop type 2 diabetes.

Smoking may also affect sexuality and the reproductive system. Smokers have a higher risk of infertility, and women who smoke pre-menopause are at an increased risk of cervical cancer and are more likely to experience pregnancy complications (e.g., miscarriages and premature deliveries).

Pregnant women who smoke may suffer complications in the birth, and newborns who inhale second-hand smoke are susceptible to the ill effects as well.

Hazardous chemicals in the environment

Over the past century, humans have introduced many chemical substances into the environment. These wastes come from many different sources, including industrial and agricultural processes. While some chemicals are very useful, many of them are toxic, and their harmful effects on the environment and human health far outweigh their societal benefit.

The balance between human activity and environmental sustainability requires attention. Human activities have a multifaceted impact on the environment and create a chain of interconnecting reactions that may harm different ecosystems.

Types of hazardous waste

A *World Health Organization* (WHO) study suggested that priority pollutants should be defined by toxicity, environmental persistence and mobility, bioaccumulation, and other hazards such as explosiveness.

Pollutants considered to have the greatest potential to impact human health based on their environmental persistence, bioaccumulation, the amount emitted, and toxic levels are cadmium, mercury, arsenic, chromium, nickel, dioxins, polychlorinated biphenyls (PCB) and sulfur dioxide.

Microbial pathogens also pose a potential threat, particularly in composting, sewage treatment, and landfill.

Organochlorines, such as PCB, are dangerous chemical compounds that were developed for use in electric equipment as cooling agents. During the manufacture and disposal of products containing PCBs, millions of gallons of PCB oil leaked out. The manufacture of PCBs ceased in the United States in the 1970s.

Despite being phased out of manufacturing, organochlorines are still present in the environment. These chemicals, which are hard to detect and almost indestructible, accumulate in the food chain for decades and can be observed in high levels years after their production has been discontinued.

Today, the chemical is still found in marine species, especially mammals and seabirds. PCBs are carcinogenic, with the potential for damaging the liver, the nervous system, and the reproductive system in adults.

Burning PCBs has resulted in the formation of dioxins, which are even more toxic.

Figure 6.17. *A worker labels PCB-containing transformer.*

Dioxins are some of the most toxic human-made organic chemicals; only radioactive waste is more toxic. They are a class of super-toxic chemicals formed as a by-product of the manufacture, molding, or burning of organic chemicals and plastics that contain chlorine. These chemicals have serious health effects even at levels as low as a few parts per trillion.

Dioxins are practically indestructible and are excreted by the body extremely slowly. These compounds enter the body in food and accumulate in body fat. They bind to cell receptors and disrupt hormone functions in the body, affecting the normal functioning of genes. The human body is not designed to defend against dioxins, which may result in a variety of problems, ranging from different types of cancers, reduced immunity in the nervous system, blood disorders, miscarriages, and congenital disabilities.

The effects vary in magnitude from very noticeable to subtle. The alteration of gene function can cause genetic diseases that may affect childhood development, result in Attention Deficit Disorder, diabetes, endometriosis, chronic fatigue syndrome, and rare nervous system or blood disorders.

Cadmium is a naturally occurring element in the Earth's crust. When combined with other elements it forms compounds such as cadmium oxide, cadmium chloride, or cadmium sulfate, all of which are highly toxic. This metal is used for its resistance to corrosion to manufacture batteries, plastics, pigments, and metal coatings.

Cadmium enters the environment through landfills, poor waste disposal methods and leaks at hazardous waste sites. It is also a waste product of mining and other industrial activities.

Cadmium particles can also enter the air from the burning of fossil fuel for energy and incineration of waste. The particles are small and can travel long distances before falling to the ground or into bodies of water. Cadmium enters the food chain from being discharged into the oceans; animals and plants intake cadmium from the soil.

Food contaminated with cadmium has the potential to irritate human digestive systems, resulting in vomiting and diarrhea. If inhaled, it can damage the human respiratory system. Over time, cadmium accumulates in the body, even if the exposure is at a low level, and it can be difficult to get rid of. Accumulated cadmium can cause kidney and bone disease.

Treatment and disposal of hazardous waste

Hazardous waste management is the collection, treatment, and disposal of waste material that can cause substantial harm to human and environment health when improperly handled. Hazardous wastes can be in a solid, liquid, sludge, or gaseous form.

They are mainly generated by chemical production, manufacturing, and other industrial activities. Inadequate storage, transportation, treatment, or improper disposal can result in the contamination of surface and groundwater.

People who live near abandoned waste disposal sites may be at higher risk of being affected. Hazardous waste generated at a site often requires transportation to an approved treatment, storage, or disposal facility.

Because of the potential threat to public safety and the environment, transport is given special attention by governmental agencies. Laws are enforced to ensure proper labeling, transporting, and tracking methods are used in the disposal of all hazardous wastes.

Hazardous waste is usually transported in specially designed tanks made of steel or aluminum alloy, carried on trucks. Though far less common, hazardous waste can also be transported via rail or air transport.

Figure 6.18. *Fluor Fernald workers pneumatically remove hazardous waste.*

The *manifest system* is employed in the United States to monitor the transportation of hazardous waste from its source to treatment or disposal facilities. This system helps to eliminate the issue of *"midnight dumping"* when waste is left in secluded areas during the night. It also provides a way to determine the type and quantity of hazardous waste being generated, as well as the recommended emergency procedures in the event of an accidental spill.

A record-keeping document, called a manifest, must be prepared by the generator of hazardous waste, such as the chemical manufacturer. The generator is responsible for the final disposal of the waste and is required to give the manifest, along with the waste itself, to a licensed waste transporter. The transporter must deliver a copy of the manifest to the recipient of the waste at an authorized disposal facility.

Every time the waste changes from one person to another along its journey, a copy of the manifest must be signed. Copies of the manifest are kept by each party involved, and additional copies are sent to the appropriate environmental agencies. The transporter is required to notify local authorities and take other appropriate actions to reduce public and environmental effects in the event of a leak during transportation.

Hazardous waste can be treated by chemical, thermal, biological, and physical methods. Chemical methods include ion exchange, oxidation, and reduction, precipitation, and neutralization.

Thermal methods used to treat hazardous waste include high-temperature incineration, which can detoxify organic waste. Thermal equipment such as a fluidized-bed incinerator, multiple-hearth furnace, rotary kiln, and liquid-injection incinerators are used to burn waste in either solid, liquid or sludge form. Thermal methods have the potential to pollute the atmosphere.

Biological methods are used to treat certain organic wastes, such as those produced by the petroleum industry. *Landfarming* is a biological method used to treat hazardous waste in which the waste is mixed with surface soil.

Organisms are sometimes added to the mixture to assist in metabolizing the waste and to add nutrients, a process called *bioremediation.* Care is taken not to grow food or forage crops on the same site. These microbes can have a stabilizing effect on hazardous wastes.

Unlike chemical, biological, and thermal methods that change the molecular structure of the waste, physical treatment methods concentrate, solidify, or reduce the volume of the waste. These processes include evaporation, sedimentation, flotation, solidification, and filtration. Solidification is done by encompassing the waste in concrete, asphalt, or plastic. Encapsulation produces a solid mass of material that is resistant to leaching. Waste can also be mixed with lime, fly ash, and water to form a solid, cement-like product.

Hazardous wastes that are not destroyed by incineration or treated by various processes need to be properly disposed of. Land disposal has proven to be the most suitable disposal site, as evidenced by landfilling and underground injection.

Temporary on-site waste storage facilities include open waste piles and ponds or lagoons. New waste piles must be carefully constructed over an impervious base and must comply with regulatory requirements like those for landfills. Open pits or holding ponds, called lagoons, are built with impervious clay soil lining and flexible membrane liners to protect groundwater.

Leachate collection systems should also be installed along with groundwater monitoring wells. Open lagoons provide no treatment of the waste except for some sedimentation, surface aeration, and evaporation of volatile organics. Accumulated sludge should be handled as hazardous and must be periodically removed.

Hazardous wastes can also be deposited in secure landfills that must have two impermeable liners and leachate collection systems. Leachate is collected and pumped to a treatment plant. To reduce the amount of leachate in the landfill and minimize the potential for environmental damage, an impermeable cap or cover is placed over a completed landfill.

A groundwater monitoring system, which includes a series of deep wells drilled in and around the site, is also implemented. These wells are routinely sampled, and the material tested to detect leaks or groundwater contamination. If a leak occurs, the wells can be pumped to intersect the polluted water and bring it to the surface for treatment.

Deep well injection (see diagram below) can be used to dispose of liquid hazardous waste. Liquid waste is pumped through a steel casing into a porous layer of limestone or sandstone. The liquid is pumped into rock pores and fissures using high pressures for permanent storage.

The injection zone is required to be located below a layer of impermeable rock or clay. This method is inexpensive and requires little or no pre-treatment of the waste.

Well injection, however, does pose a danger of leaking hazardous waste and polluting subsurface water supplies.

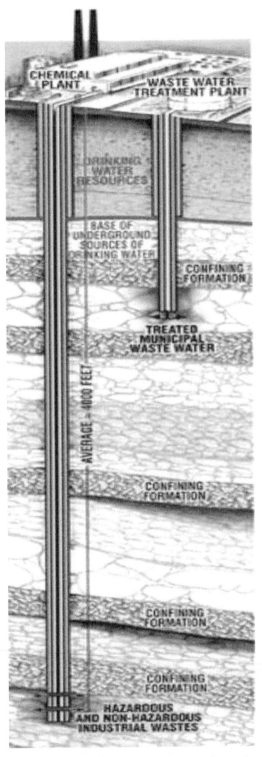

Figure 6.19. *Deep injection well for disposal of hazardous, industrial and municipal wastewater*

Cleanup of contaminated sites

Many abandoned hazard disposal sites such as pits, ponds, or lagoons are unrestricted and unlined, posing a threat to human and environmental health. Depending on the level of risk these sites pose, it may be necessary to conduct remediation.

Remediation may require the complete removal of waste material from the site and transporting it to another location for treatment and proper disposal. Alternately, remediation can be done on-site to reduce the production of leachate and decrease the chance of groundwater contamination.

On-site remediation may include the temporary removal of the hazardous waste, construction of a secure landfill on the same site, and the proper replacement of the waste. Contaminated soil or groundwater may also require treatment. Treated soil may be replaced on-site, and treated groundwater returned to the aquifer by deep-well injection.

A less costly method is the full containment of the waste. This can be done by placing an impermeable cover over the disposal site and blocking the lateral flow of groundwater with subsurface cutoff walls. These walls are required to be constructed around the entire perimeter of the site and be deep enough to penetrate to the impervious layer. They can be excavated as trenches around the site, which are filled with a bentonite clay slurry to prevent their collapse during construction.

The trenches should then be backfilled with a mixture of soil and cement to form an impermeable barrier. Cutoff walls function as vertical barriers to the flow of water, and the impervious layer serves as a barrier at the bottom.

Biomagnification

Biological magnification is the increased concentration of toxic substances in the tissues of organisms at successively higher levels of the food chain. This occurs when hazardous waste from industrial, agricultural, and human activities contaminate bodies of water, such as rivers and are consumed by animals in the food chain.

As these animals consume more of the contaminated food, the toxins become more concentrated in the animal's tissues. These pollutants cause genetic mutation, diseases, congenital disabilities, reproductive difficulties, behavioral changes, and death in aquatic organisms.

Many of these toxins settle on the seafloor and are consumed by organisms that feed on food and sediments at the bottom of the ocean. These compounds are not digested and accumulate within the animals that consume them, becoming more concentrated as the compounds move along the food chain.

When the toxin reaches the top of the food chain, it is more concentrated and more dangerous. Humans, the species at the top of the food chain, then eat contaminated animals and ingest the toxins contained within them, which can have a serious health impact.

In northwest Ontario, Canada, between 1962 and 1970, biomagnification occurred as a result of a paper plant dumping ten tons of mercury into a local river. The result was that approximately 100 residents developed Minamata disease, or methylmercury poisoning, the symptoms of which include slurred speech, tunnel vision, spasms, and may lead to insanity and eventual death. The disease is named for its city of origin, Minamata, Japan, where residents started exhibiting such symptoms after a corporation dumped its mercury waste into a bay and contaminating the fish.

Relevant laws

The *Toxic Substances Control Act* (TSCA) is a United States law, passed by the United States Congress in 1976 and administered by the EPA. This law regulates the introduction of new or already existing chemicals. The main objectives of this act are to assess and regulate new commercial chemicals before they enter the market, regulate chemicals existing in the market that posed an adverse risk to health or the environment and regulate the distribution and use of these harmful chemicals. The act specifically regulates polychlorinated biphenyl (PCB) products.

TSCA does not separate chemicals into categories of toxic and non-toxic. The act instead prohibits the manufacture or import of chemicals that are not on the TSCA. Inventory manufacturers are required to submit a pre-manufacturing notification to the EPA before they manufacture or import new chemicals for commercial purposes. The agency reviews these notifications, and if a proposed chemical is believed to have the potential to harm human health or the environment, its use is banned.

The *Chemical Safety Improvement Act* (CSIA) was bipartisan legislation proposed in 2013. It was introduced to Congress as an attempted means to reform and modernize the TSCA. This act regulates the introduction of new or already existing chemicals.

Notes

Economic Impacts

Pollution becomes an expense when it costs money to solve the problems caused by its effects. For example, a factory may pollute a river during production. If that river is used as a tourist attraction, revenue will be lost from its contamination. The city may have to absorb the cost to remediate the damages. This is referred to as an external cost because the factory is not held responsible for its actions to produce goods.

Figure 6.20. *Debris and injured birds caused by water pollution.*

Industrial waste has been discarded into the environment for over 150 years without any regard for the environment or the economy. The initial view was that the environment had an infinite capacity to absorb pollutants. It was then unknown that there are thresholds beyond which levels for different pollutants become lethal.

Before these thresholds are reached, biological damage is being done, and ecological costs are being incurred that are not immediately observable. These negative effects will eventually be translated into economic costs once their effect becomes more readily apparent.

The recent, enthusiastic interest in controlling pollution is due to many of these thresholds being reached and the beginning effect on the economy. Research has revealed that approximately 25% of respiratory disease in the United States is the result of air pollution. According to the *Organization for Economic Co-operation and Development*

(OECD), the cost of air pollution in deaths and health effects of just OECD countries, plus China and India, is 3.5 trillion dollars per year.

Cost-benefit analysis

The concept of scarcity is used by environmental economists to carry out cost-benefit analyses to recommend pollution policies to be implemented. This study assesses how people make decisions when they are faced with scarcity. Humans have infinite wants but live in a world with finite resources.

Scarcity implies that resources devoted to one end are not available to meet another. For instance, funds used by a municipality to modify its water treatment plant to remove trace amounts of arsenic cannot also be used to improve local elementary education.

Scarcity demands compromises, so some may see the total elimination of pollution as undesirable and unnecessary. The subsequent response is to manage instead and eliminate *some* pollution. Economists use the cost-benefit analysis to decide the best options for what pollution sources to eliminate.

Cost-benefit analysis provides an organizational framework for identifying, quantifying, and comparing the costs and benefits in the monetary value of proposed policy action. The final decision is based on the comparative total costs and benefits. The benefits of environmental regulations can include a decrease in human and wildlife mortality, improved water quality, species preservation, and greater recreation opportunities. The costs are usually reflected in higher prices for consumer goods and higher taxes.

While it is easy to put a dollar value on the market effects, such as improving education, it proves more difficult to attach a cost to non-market effects, such as decreasing pollution and its effects on ecosystems. This analytical tool shows how the interests, claims, and opinions of parties affected by a proposed regulation can be examined and compared. It demands specificity and practicality that is necessary for making sound decisions.

Externalities

An *externality* is a cost or benefit that affects a party who did not choose to incur that cost or benefit. If external costs (i.e., pollution) exist, the producer may choose to make more products than would be made if the producer were required to pay all associated environmental costs. Responsibility for the consequences of an action is not solely internal; an element of externalization is involved.

If there are external benefits to be gained, such as public safety, less of the good may be made than if the producer had received payment for the external benefits to others. It can be deduced that the overall cost and benefit to society is the sum of the imputed monetary value of benefits and costs to all parties involved.

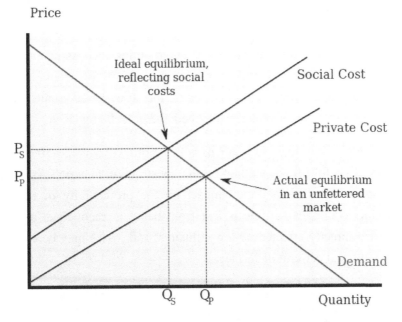

Figure 6.21. *Demand curve showing the microeconomic concept of a negative externality*

A voluntary exchange may reduce the well-being of society if external costs exist. The person who is affected by the negative externalities in the case of air pollution will see it as a lowered utility: either personal displeasure or the potential to incur costs, such as higher medical expenses. The externality may even be viewed as a trespass on their lungs, which violates their property rights.

Therefore, an external cost has the potential to pose an ethical or political problem. Otherwise, it might be seen as a case of poorly defined property rights, as with the pollution of bodies of water that may belong to no one or are publicly owned.

A positive externality, on the other hand, would increase the utility of third parties at no cost to them. Since collective societal welfare is improved, but the providers have no way of monetizing the benefit, smaller quantities of the goodwill be made than would be optimal for society.

There are numerous methods of improving overall social utility when externalities are involved. The market-driven approach to improving externalities is to "internalize" third party costs and benefits (e.g., requiring a polluter to repair any damage caused). However, in many instances, internalizing costs or benefits is impractical, especially if it is impossible to determine the true monetary value.

Many negative externalities are related to the environmental consequences of production and use. Externalities of air pollution include the economic cost, the social cost, and the environmental cost.

Economic costs include a variety of externalities such as damage to property, structures, and infrastructure, as well as loss of productivity of people and crops. Acid deposition, smog, and ozone pollution alter the time-based scale during which investments on infrastructure can be remunerated and must be replaced.

For example, buildings that are usually amortized over 20-30 years may lose from 1-5 years of useful life, depending on the construction material. Aside from health costs, air pollution also has a direct impact on the productivity of the labor force. Productivity time decreases as time is lost from being in recovery. Crops and timber products are also directly affected by air pollutants, and losses may be calculated based on quantities produced per unit of surface.

Figure 6.22. *Smog-damaged plant at the statewide air pollution research center, University of California. The plant was damaged by fumigation with 1/2 part per million of ozone for 3 hours.*

Social costs are related to the physiological impact that pollution has on humans, especially on the cardiovascular and respiratory systems. Some impacts are obvious, such as the effects of carbon monoxide. Other effects are less discernible and more malignant, such as lead and VOCs. It would be difficult to attribute a specific case of lung cancer to general air pollution or a smoking habit. This is because a large percentage of the population reside in urban areas and are constantly exposed to air pollution emissions.

Environmental costs encompass overall damage done to the ecosystem through the atmosphere, except for what may be considered economically useful to human activities such as agriculture. Environmental costs are the most difficult to assess comprehensively. They can encompass a wide variety of effects, such as biological diversity and sustainability, which may prove difficult to quantify.

Marginal costs

Marginal cost is an economic term that refers to the change in total production cost brought about by the production of an additional unit. For example, if a factory chooses to produce a new line of a certain product, the marginal cost of the new product line would include all the additional costs such as extra materials, added production, and additional work hours.

There is an optimum amount of pollution where marginal benefit equals the marginal cost of pollution. To determine the optimum amount of pollution, taken into consideration are marginal cost, marginal abatement cost, and marginal benefit.

The economics of pollution suggests that there is an "ideal" amount of pollution where the wellbeing of society is at its peak, and the damage to the environment is at a sustainable level. Economists have argued that it is not economically or environmentally wise to eliminate pollution. This is because the cost to reduce pollution reduction would likely exceed the benefits.

Aquatic environments and the atmosphere have a natural ability to assimilate some amount of pollutants with no ill effect on the environment or humans. To disregard this, natural assimilative capacity would be wasteful. Furthermore, what one industry regards as a pollutant may be another industry's raw material. Marginal cost, for pollution, involves an additional environmental cost that results from the production of an additional unit. Therefore, an additional environmental cost may result in an increase in greenhouse gas emissions that arise from the production of the new product line.

In addressing the issue of the optimum amount of pollution, the marginal abatement cost must be analyzed. This cost refers to the expense incurred by a company to remediate a harmful situation it created. This cost is therefore associated with eliminating a unit of pollution. For an example of an *abatement cost*, consider a coal-burning power plant that produces toxic sulfur dioxide emissions, which have detrimental environmental effects.

To reduce this problem, the company may install smokestack scrubber technology to remove these emissions. The cost associated with installing and using the scrubbers is an abatement cost. As the amount of pollution released decreases, the marginal abatement cost increases. This is because more efficient and advanced technologies used to control pollution are more costly.

The point where society's well-being is at its optimum, concerning environmental quality, is when the marginal cost of pollution abatement is exactly equal to the marginal benefit from pollution abatement. If the marginal benefit of reducing pollution exceeds the marginal cost of reducing pollution, then society would benefit from a reduction in pollution.

The benefit would be equal to the amount by which the marginal benefit of the cleanup exceeded the marginal cost of cleaning the environment. However, if the marginal cost of pollution abatement exceeds the marginal benefit from the reduction, then the benefit of cleaning the environment is not worth the expense; further attempts to clean up the environment will result in a reduction in welfare.

Sustainability

Sustainability refers to the ability to support current needs without compromising future supply. *Environmental sustainability* involves preserving the environment to support long-term ecological balance. Pollution prevention is an important part of sustainability and long-term planning.

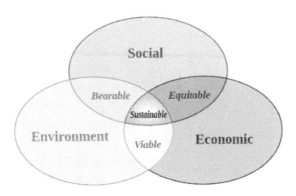

Figure 6.23. *Sustainable development.*

Focusing on pollution control and environmental sustainability has a significant impact on solving pollution problems.

Sustainable Business Practices (SBPs) involve designing businesses in a manner to reduce or eliminate dependence on finite resources. This includes activities such as conserving areas around cities and using sustainable energy sources like solar, wind, geothermal, or biofuels instead of fossil fuels. Also, it involves designing products that are recyclable and encouraging people to buy recycled products.

The principles of sustainability apply to all aspects of business, such as construction, development, and production. Pollution prevention is the first step in achieving sustainability. An understanding of how waste is generated and how it can be minimized is the first step to eliminating waste, increasing efficiency, and developing sustainable production methods.

Pollution prevention has both economic and environmental benefits, including minimizing the amount of energy and resources used to treat and dispose of waste. It also ensures that waste produced can be reused or disposed of easily and without harming the environment. In this way, pollution prevention becomes an important aspect of sustainability as the cost for extra raw materials, waste disposal, and waste treatment systems can be eliminated or substantially reduced.

New tools and processes to prevent pollution at its source and institute sustainability practices are becoming more popular. Business models of companies can be reviewed to identify opportunities to incorporate pollution prevention practices.

Notes

Chapter 7

Global Change

Stratospheric Ozone
- Formation of stratospheric ozone
- Ultraviolet radiation
- Causes of ozone depletion
- Effects of ozone depletion
- Strategies for reducing ozone depletion
- Relevant laws and treaties

Global Warming
- Greenhouse gases and the greenhouse effect
- Impacts and consequences of global warming
- Reducing climate change
- Relevant laws and treaties

Loss of Biodiversity
- Habitat loss
- Overuse
- Pollution
- Introduced species
- Endangered and extinct species
- Maintenance through conservation
- Relevant laws and treaties

Stratospheric Ozone

The Earth's stratospheric ozone layer contains approximately 90% of all the ozone in the atmosphere. This ozone is what makes the Earth habitable, as it absorbs the harmful ultraviolet radiation from solar rays before they can reach the Earth's surface. The UV radiation is harmful and causes sunburn and damage to cells, which can accelerate the aging of the skin and lead to cancer.

Ozone is produced through various processes in the stratosphere, where it aids in absorbing about 99% of the harmful UV rays coming from the sun. Ozone is also produced near the Earth's surface in a layer called the troposphere, where it can be harmful. The formation of stratospheric ozone occurs through a process called the photolysis of O_2 (oxygen gas).

Photolysis of O_2 means "the breaking" (lysis) by "light" (photo). This process does not occur in the troposphere because the ozone has completely absorbed the strong UV rays needed to break the oxygen gas into smaller components in the stratosphere.

Breathing ozone can cause many health problems (e.g., inflaming and damaging nasal passages, shortness of breath, and pain when inhaling) and make the lungs more susceptible to infection. Fortunately, most of the ozone resides in the stratosphere and is not inhaled by life forms on Earth.

Formation of stratospheric ozone

Stratospheric ozone is formed naturally in the atmosphere by several chemical reactions that involve oxygen molecules and solar UV radiation (sunlight).

First, UV radiation breaks apart one molecule of oxygen gas (O_2) to produce two atoms of oxygen (2 O).

After oxygen gas is split, each of the oxygen atoms becomes highly reactive and combines with a molecule of oxygen gas (O_2) to form a molecule of ozone (O_3).

If UV radiation is present in the stratosphere, these reactions continue.

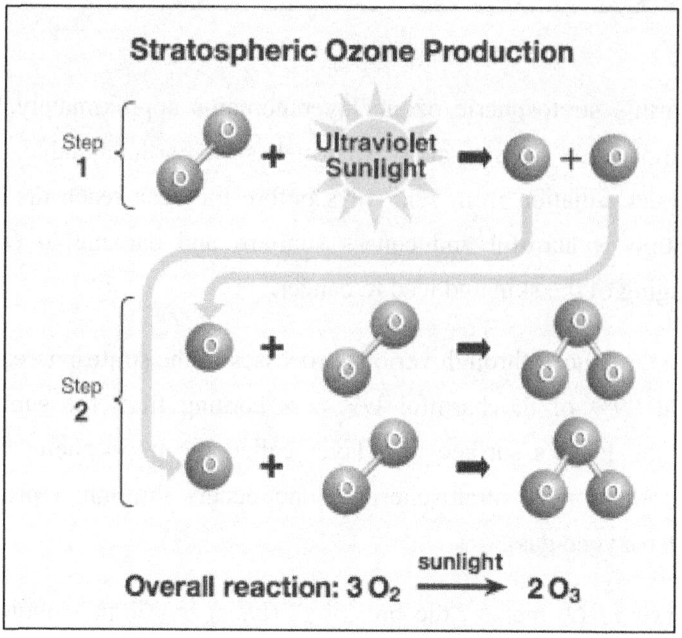

Figure 7.1. *Stratospheric ozone production.*

Ultraviolet radiation

Ultraviolet radiation (UV) comes from the sun, and it has always been an important aspect of human life and had an important role in our environment.

However, UV radiation can occur at varying wavelengths, which have different effects; to benefit from the helpful effects of UV radiation, the harmful ones must also be dealt with.

The longest UV radiation wavelengths occur at 320-400 nm and are called *UV-A* wavelengths. UV-A plays an essential role in the formation of Vitamin D by the skin, but it can also be harmful as it causes cataracts in the eyes and sunburn of the skin.

Another form of UV radiation is termed *UV-B* and occurs at shorter wavelengths of light (290-320 nm). UV-B is harmful at the molecular level in cells, and it causes damage to the fundamental building blocks of life, deoxyribonucleic acid (DNA).

DNA readily absorbs the UV radiation, which in turn causes the shape of the DNA molecule to change. Changes in their shape will prohibit protein-building enzymes from attaching to the strand and properly reading the DNA molecule. When DNA is not read properly, distorted proteins may be created, or cells may die from the high level of dysfunction.

Types of radiation

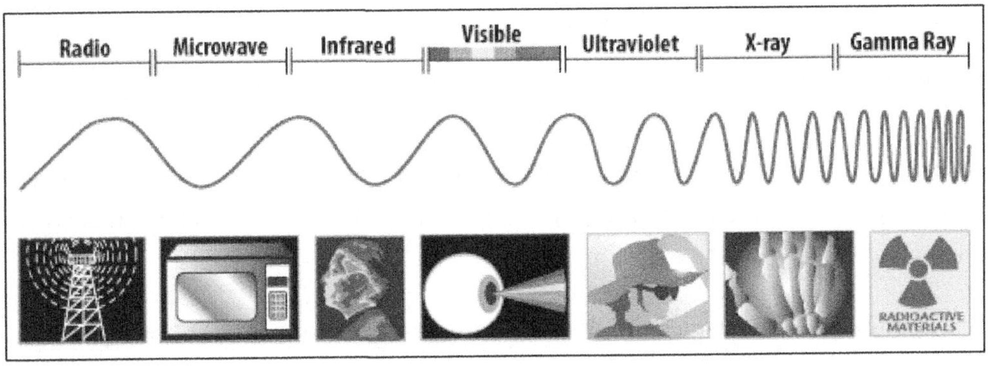

Radio	Microwave	Infrared	Visible	Ultraviolet	X-ray	Gamma Ray

Figure 7.2. *Various types of radiation that compose the electromagnetic spectrum.*

Though radiation is often thought to be bad, some forms of radiation are beneficial and important parts of everyday life. Infrared radiation is energy that is given off by the Earth's surface to cool it down.

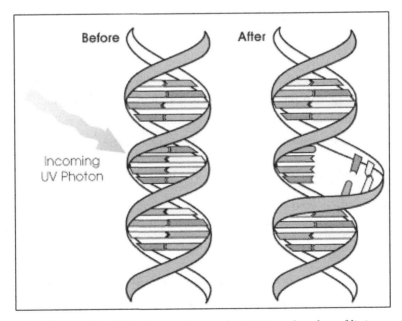

Figure 7.3. *Ultraviolet (UV) photons harm the DNA molecules of living organism*

In one common damage event, adjacent bases bond with each other instead of across the strands to their complementary base. This creates a bulge in the DNA double helix, causing the DNA not to function properly.

Causes of ozone depletion

There is scientific evidence that chemicals in the atmosphere containing bromine or chlorine are depleting stratospheric ozone. These are *"ozone-depleting substances"* (ODS). These chemicals are very stable, safe for the environment in the lower atmosphere, and generally not toxic. Their molecular stability allows them to float up into the atmosphere and interact with UV light rays, where they have broken apart and release bromine or chlorine.

Both elements deplete ozone at an astonishingly fast rate by stripping an atom of oxygen from the ozone molecule. Just one molecule of bromine or chlorine can tear apart thousands of molecules of stratospheric ozone. In addition to ODSs' ability to destroy ozone molecules, they have a very long lifetime in the atmosphere.

It can take several centuries for them to dissipate; the ODS substances released by human activities 70-100 years ago are still working their way up to the stratosphere.

The main ODS that cause ozone depletion:

- *Chlorofluorocarbons* (CFC) are the most widely used and account for more than 75% of the stratospheric ozone depletion. They were found in freezers, air conditioners, and refrigerators as coolants, as well as in foam products that provide padding in shipping containers.

- *Halons* contain bromine and are used in fire extinguishers.

- *Hydrofluorocarbons* became a substitute for CFCs after the damage CFCs cause became more well-known. While they are not as harmful to stratospheric ozone as CFCs are, they still cause ozone destruction.

- *Methyl Chloroform* is utilized mainly in the industry for products such as aerosols, adhesives, and the processing of various chemicals.

- *Carbon tetrachloride* is used in fire extinguishers and industrial solvents.

Effects of ozone depletion

Stratospheric ozone works as a filter; it catches most of the sun's shortwave UV radiation. If the ozone layer becomes depleted, more ultraviolet rays will reach the Earth. Exposure to high levels of UV radiation can be detrimental not only to humans but also to plants and animals.

Human health will be affected in several ways, including:

- Increased number of skin cancers, often caused by exposure to UV rays that burn the skin;

- Premature skin aging;

- Immunosuppression, or weakened immune system (recent studies have shown that too much exposure to UV radiation can affect the efficacy of the human immune system); and

- Blindness, cataracts, and various other eye diseases (UV radiation had been shown to damage parts of the eye, such as the retina, the cornea, the lens, and the conjunctiva).

Agriculture, ecosystems, and forests will be affected because many major crop species are harmed when exposed to increased UV radiation. This decreases the overall growth and photosynthetic processes in the plants, which decreases crop production. Species that are sensitive to UV radiation levels include corn, rice, oats, barley, wheat, tomatoes, cucumbers, carrots, peas, and broccoli.

Animals will no doubt be affected; they too, can experience skin and eye cancers from high levels of UV radiation. In the Antarctic ozone hole, UV levels have already affected the developmental stages of various aquatic life forms.

Marine life will experience damage, in particular, plankton. Plankton are tiny life forms that reside in the first few meters under the surface of the ocean are the first step in many aquatic food chains. When the number of plankton decreases, the food chain is disrupted, and there is a loss of biodiversity in our oceans. This will, in turn, affect humans, many of whom are wholly or partly reliant on aquatic life forms as a food source.

Wood, rubber, plastic, and fabrics are degraded by UV radiation. Protecting and replacing these materials would be a heavy financial burden.

Strategies for reducing ozone depletion

While the EPA and other agencies around the globe monitor and enforce laws that help reduce the amount of ODS in the atmosphere, there are actions that can also be taken on a smaller, more personal level to help, such as:

- *Limiting driving of private vehicles.* Emissions from automobiles greatly add to the amount of smog in the atmosphere, a major culprit in the depletion of the ozone layer. Alternatives to driving a personal vehicle include public transportation, walking, riding a bicycle, and getting a ride with a friend. There have been advances in personal transport in the past few decades, such as electric and hybrid vehicles, that aid in the reduction of emissions;

- *Avoiding the use of pesticide-containing products.* Though they are an easy solution for ridding your garden of weeds, their effect on the atmosphere is long-lasting. Natural remedies for weed removal or removal by hand are two alternatives to using pesticides; and

- *Using eco-friendly household cleaning.* Many cleaning products that are readily available for public use contain toxic chemicals that may harm your health if not handled properly and expedite ozone depletion. Natural cleaning products are now sold in almost every supermarket and health store.

Relevant laws and treaties

Many strategies can be implemented to reduce the depletion of ozone in the atmosphere. In 1987, 194 nations signed an agreement to end all production of chlorofluorocarbons, halons, and various other substances that are known to deplete ozone (ODSs); this agreement is the *Montreal Protocol on Substances that Deplete the Ozone Layer.*

Over the years, it has been amended to help quicken the phase-out of ODSs. Each country has undertaken its efforts to reduce ozone depletion. The EPA is the federal agency in the United States, primarily responsible for air quality management and the protection of the atmosphere.

Under the Clean Air Act, countries have created regulatory programs to address various problems, including:

- Ending the production of *ozone-depleting substances* (ODSs);

- Ensuring that refrigerants and halon fire extinguishing agents are recycled properly;

- Identifying safe and effective alternatives to ozone-depleting substances;

- Banning the release of ozone-depleting refrigerants during the service, maintenance, and disposal of air conditioners and other refrigeration equipment; and

- Requiring that manufacturers put warning labels on products either containing or made with the most harmful ODSs.

In addition to enforcing the Clean Air Act, the EPA also works with other federal and international agencies toward the goal of finding new methods of cleaning up the stratosphere. Both the *World Meteorological Organization* and the *United Nations Environment Programme* gather scientific evidence for the laws that will be enacted.

To help the American people protect themselves from harmful exposure to UV radiation, the EPA has put several educational services in place to inform the public of the dangers of unnecessary exposure to UV radiation. They are partnered with the National Weather Service so that daily weather forecasts contain a UV Index, indicating the intensity of UV radiation for the days ahead.

The EPA also has a program called the SunWise School Program, which educates children on the effects of overexposure to sunlight, as well as practical steps they can take to keep themselves safe.

Also, the products that are combusted during the trip of a spacecraft into outer space are very harmful to the ozone layer. These products do not take tens or hundreds of years to reach the stratosphere, as spacecraft travel through it and expel ODS directly into the middle and upper layers. While regulating rocket launches may not seem like a viable initiative, scientific advances and EPA regulations are changing the way NASA and other space programs operate.

Global Warming

Global warming is defined as a gradual increase in the average temperature of the Earth's oceans and atmosphere. Over the past several decades, this increase has been exacerbated by emissions from automobiles, factories, and other machines. Scientists believe this change will permanently alter the Earth's climate and consider it the greatest environmental threat of our time.

Climate change is a term often used to refer to the same phenomenon of increasing global temperature, but it can also refer to climate trends at any point of the Earth's history. Global warming focuses more on the current warming of the Earth's climate.

In 2014, the *Intergovernmental Panel on Climate Change* (IPCC) stated that scientists are over 95% sure that global warming is being caused by an increase in greenhouse gases and anthropogenic forces.

Anthropogenic forces are any occurrences that take place due to human activity. Scientists commonly use a 5% margin of error when determining how certain they are about a conclusion.

When a researcher concludes in a study that they are 95% confident that some action causes a response (such as human activity causing global warming), it means that there is a 5% chance or less that there is an error in their statement. Generally, scientists aim for a 95% confidence level or higher when making statements about their research findings because it allows for very little error.

When scientists state that they are 95% certain that greenhouse gases and human activity are causing global warming, they are asserting that their research is highly supportive of this statement.

Why are seemingly minuscule changes and variations in temperatures concerning? Of all the planets in the solar system, Earth is the only one that houses life forms that require a certain range of temperatures. This range is balanced by solar heat from the sun and thermal energy that is radiated back out to space.

Oceans, plants, storms, the sea level, and the concentration of salt in the oceans are all influenced by the Earth's surface temperature. Even the smallest variation in greenhouse gases can cause catastrophic changes.

Greenhouse gases and the greenhouse effect

Terminology:

- *Average Surface Temperature* is the average of the Earth's sea-surface and above-ground temperatures. This value is calculated by taking the average temperature of water of the first few feet below the surface of the ocean and the temperature found between the surface of the Earth and 1.5 meters above the surface.

- *Infrared radiation* is a form of energy given off by the Earth to cool down from heat absorbed by solar energy. This solar energy, called sunlight, is both absorbed and reflected by the surface of the Earth.

- A *sink* is a reservoir that absorbs an element or compound as a part of its natural cycle.

- A *source* is any process through which greenhouse gas is emitted into the atmosphere.

- *The Nitrogen Cycle* is a natural circulation of nitrogen molecules among plants, animals, microorganisms, and the atmosphere. The molecules of nitrogen can take on several different forms, depending on what stage they are in and how they are needed for a specific process. Bacteria break down nitrogen in the ocean and in the soil, which naturally releases nitrous oxide, a greenhouse gas. Nitrous oxide can be removed from the atmosphere by way of chemical reactions, UV radiation, and absorption by certain forms of bacteria.

- *Global Warming Potential* (GWP) measures how much energy the emission of one metric ton of a given gas will absorb over a certain period, about the emissions from one metric ton of carbon dioxide. The greater the GWP, the more that gas will warm up the Earth compared to carbon dioxide over a given period. For instance, carbon dioxide has a GWP of 1, no matter the time frame being used and is always the gas in reference. Nitrous oxide is believed to have a GWP of 265-298 times the GWP of carbon dioxide over 100 years. Also, fluoridated gases possess GWP levels that are thousands or tens of thousands of times higher than the GWP levels of carbon dioxide.

- The *greenhouse effect* is when greenhouse gasses trap infrared radiation before it can escape into outer space. This warms the atmosphere and eventually causes the surface of the Earth to warm up as well. This process got its name from a similar process that is seen on a smaller scale within an actual greenhouse. A greenhouse is a structure made of glass that lets sunlight enter and prevents heat from escaping. This allows the inside of the building to remain warm even though the temperature outside is too cool to sustain certain plant life.

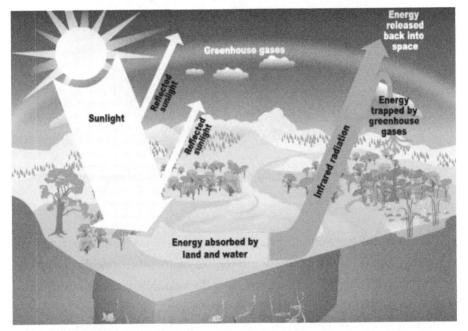

Figure 7.4. *A condensed cycle outlining the greenhouse effect.*

Most scientists agree that the main cause of the global warming phenomenon is the greenhouse effect, which disrupts the stability of the various layers of the atmosphere through which infrared radiation passes. There are several gases in the atmosphere that can block heat from escaping, which in turn increases the Earth's average temperature.

Average surface temperature and average atmospheric temperature are two measurements used for analyzing trends. The Earth's average annual surface temperature is much higher today than it was when it was first recorded in the mid- to late 1880s. All ten of the warmest years on record have taken place since the last El Niño event in 1998.

Gases that contribute to causing the greenhouse effect include fluorinated gases, nitrous oxide, carbon dioxide, and methane.

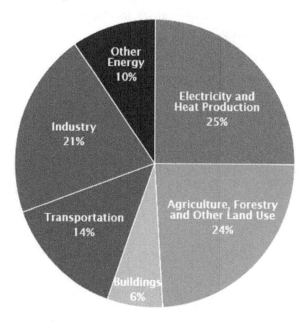

Figure 7.5. *Worldwide greenhouse gas emissions by economic sector.*

Fluorinated Gases

Often called F-gases, fluorinated gases only come from anthropogenic forces and have no natural source on the planet Earth (unlike many other greenhouse gases). There are four main categories of F-gases: nitrogen trifluoride (NF_3), sulfur hexafluoride (SF_6), perfluorocarbons (PFCs), and hydrofluorocarbons (HFCs). They are released through many industrial processes such as semiconductor and aluminum production.

In comparison to other greenhouse gases, F-gases have a very, very high global warming potential (GWP), so even small concentrations can have immense effects on the average global temperature. They have lifetimes in the atmosphere of up to several thousand years and can only be destroyed by solar radiation in the upper atmosphere.

- *Substitution for ozone-depleting substances* accounts for approximately 90% of all F-gases emitted in the United States. Hydrofluorocarbons are used as fire retardants, refrigerants, solvents for use in laboratories, and aerosol propellants. Because F-gases do not deplete the stratospheric ozone layer, they were manufactured as a replacement for those chemicals that did—chlorofluorocarbons and hydrochlorofluorocarbons.

- *Transmission and distribution of electricity* involve the use of sulfur hexafluoride in types of electrical transmission equipment such as circuit breakers. The Intergovernmental Panel on Climate Change has stated that sulfur hexafluoride is the most potent of all greenhouse gases, with a GWP of 22,800.

- *Industrial processes* produce perfluorocarbons as a by-product in semiconductor and aluminum production/manufacturing. PFCs also have a high GWP and a long atmospheric lifetime, though they are not considered as potent as sulfur hexafluoride.

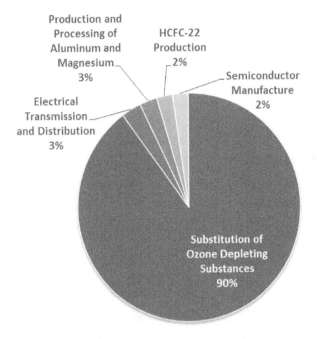

Figure 7.6. *Fluorinated gas emissions in the United States, by source. All emissions estimates are from the Inventory Greenhouse Gas Emissions and Sinks: 1990-2013.*

Nitrous Oxide

Nitrous oxide (N_2O) accounts for approximately 5% of all greenhouse gas emissions from human activity in the United States. It occurs commonly throughout nature, and it is a part of the Earth's nitrogen cycle. Molecules of nitrous oxide stay in the atmosphere for about 114 years before being destroyed through a series of chemical reactions or being removed by a *sink*. The impact of nitrous oxide on the atmosphere is huge.

Scientists estimate that the impact of one pound of N_2O on the warming of the atmosphere is nearly 300 times the impact of one pound of CO_2. Human activity through agriculture, wastewater management, factories, and the combustion of fossil fuels has led to a great increase in the amount of N_2O in the atmosphere.

- *Transportation* adds to the amount of nitrous oxide in the atmosphere. Automobiles emit N_2O as a byproduct when fuel is burned. Motor vehicles such as trucks and cars produce the most nitrous oxide of all vehicles. As with carbon dioxide, the amount of N_2O released into the atmosphere depends on the type of fuel being burned, as well as the vehicle itself.

- *Industrial processes* emit nitrous oxide into the atmosphere through the production of nitric acid; an ingredient used to make commercial synthetic fertilizer. It is also emitted during the production of adipic acid, used to make various fibers such as nylon, as well as other synthetic products.

- *Agriculture* soil management is the largest contributor to nitrous oxide emissions in the United States, accounting for nearly 75% of total U.S. nitrous oxide emissions in 2013. Current soil management practice involves adding nitrogen to the soil to produce a nitrogen-rich fertilizer. This process emits nitrous oxide as a byproduct, which then is absorbed into the atmosphere.

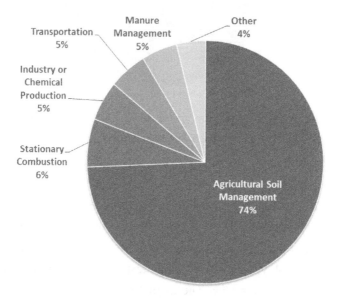

Figure 7.7. *Nitrous Oxide (N_2O) emissions in the United States by source. All emissions estimates are from the Inventory Greenhouse Gas Emissions and Sinks: 1990-2013.*

Carbon Dioxide

Carbon dioxide (CO_2) accounts for approximately 82% of all greenhouse gas emissions in the United States, and it is the primary GHG emitted by anthropogenic forces. It is released through natural processes such as human (and animal) respiration, deforestation, volcanic eruptions, and the burning of fossil fuels. Carbon dioxide is constantly cycling through the ocean, the atmosphere, and various land surfaces because it is produced by plants, animals, and microorganisms such as bacteria. When nature is left to produce and consume CO_2, the levels in the atmosphere level out well and remain in a good balance.

Since the onset of the Industrial Revolution in the 1750s, human activity has increased atmospheric CO_2 concentration by one-third and added substantially to climate change. Human activities have begun to change the carbon cycle by adding more carbon dioxide into the air and changing the ability of forests to remove CO_2 from the atmosphere (deforestation decreases the number of trees available to absorb the carbon dioxide, leaving more CO_2 present in the atmosphere).

The main human activity that releases carbon dioxide into the atmosphere is the combustion of fossil fuels used to generate electricity and power automobiles, trucks, trains, and airplanes.

- *Transportation* is a large contributor to carbon dioxide emissions; CO_2 is released with the burning of fossil fuels such as gasoline and diesel. This includes methods of transportations such as air travel, highway automobiles, railroads, and marine transportation such as cargo ships and cruisers.

- *Industrial processes* such as the production and consumption of cement and metals such as steel and iron and the production of chemicals all emit carbon dioxide through the combustion of fossil fuels. Many industries require immense amounts of electrical power to operate and indirectly burn more fossil fuels due to the burning of hydrocarbons.

- *Electricity* use and generation account for the largest portion of carbon dioxide emissions in the United States. Fossil fuels are combusted to produce electricity. Although other methods are in use today, such as hydro and wind energy, fossil fuels remain the main source of energy that most countries depend on as a source for generating electricity.

The amount of carbon dioxide released into the atmosphere depends on the type of fossil fuel being combusted. The burning of coal, for example, will emit more carbon dioxide than will the burning of natural gas or oil.

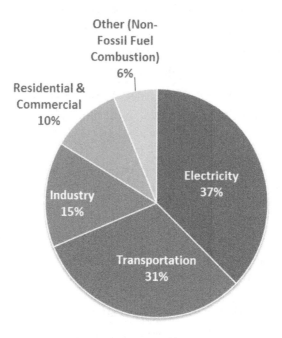

Figure 7.8. *United States carbon dioxide (CO_2) emissions, by source. All emissions estimates are from the Inventory Greenhouse Gas Emissions and Sinks: 1990-2013.*

Methane

Methane (CH_4) accounts for around 10% of all greenhouse gas emitted in the U.S. due to human action, as opposed to natural processes. It is emitted from human activities such as natural gas system leaks and the raising of livestock, as well as from areas such as wetlands.

Though methane does not last very long in the atmosphere (12 years), it is much more efficient at trapping infrared radiation than carbon dioxide is.

Natural processes that occur during chemical reactions in the atmosphere help remove methane from the air. Globally, methane emitted from human activities makes up 60% of all methane emissions. It is released through agricultural practices, various industrial practices, and waste management activities.

- *Wastes* from businesses and houses accrue in landfills around the country. As the wastes decompose, methane gas is released. Landfills are the third-largest contributor to methane emissions in the U.S.

- *Agriculture* contributes to methane emissions, mostly due to the raising of livestock such as cows, sheep, goats, and buffalo. It is released by the animals as a part of their digestive processes and is also emitted from manure that is stored for later use as fertilizer. Globally, agriculture accounts for the largest source of methane emissions.

- *Industries* contribute to methane emissions, mainly by their natural gas and petroleum systems. Methane is the primary component of natural gas that is drilled from underground sources, and it can easily be released during the various stages of processing, production, storage, transportation, and distribution.

Methane is often found underground alongside crude oil reserves, so it will also be found in the petroleum that is extracted. During the processing and production of crude oil, methane is often released into the atmosphere.

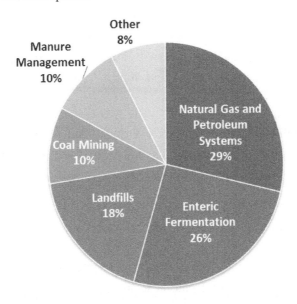

Figure 7.9. *United States methane emissions (CH$_4$), by source. All emissions estimates are from the Inventory Greenhouse Gas Emissions and Sinks: 1990-2013.*

Impacts and consequences of global warming

Climate change can have several effects on the planet Earth, such as influencing crop yields, affecting human health, causing changes in rainfall, altering forests and ecosystems, and changing the available supply of energy.

Impacts of global warming vary in different regions and ecosystems across the planet. For instance, a desert in Africa will experience a slight decrease in rainfall differently than the rainforests of South America.

Outlined below are various sectors of life on planet Earth that have been and will continue to be affected by global warming:

- *Energy* demands will change with fluctuating temperatures, increased demands for cooling during the summer, and an increased need for heating during the winter. Frequent and intense storms, along with the continuing rise in sea level, may disrupt the extraction and transportation of fuels (e.g., drilling rigs on the ocean);

- *Agriculture* (e.g., crop yields) may be affected by drought, flooding, and severe warming. While slight changes in the amounts of carbon dioxide in the atmosphere may help plant growth (it will provide more "food" for plants to metabolize), the high levels that are projected over the next several decades could prove catastrophic;

- *Transportation* infrastructure will be affected through higher temperatures, severe storms and higher storm surges along the coasts;

- *Coasts* will experience more stress during global warming; they have already been affected by human activity. Increased ocean acidity and temperature will no doubt change coastal and marine ecosystems;

- *Ecosystems* could be fundamentally transformed by climate change; it can change where species live and how they interact.

  For instance, if flooding in a certain area damages a species' habitat, it may cause them to relocate. The species instinctively looks for a new environment and could begin adapting to one that is less ideal for them than their previous habitat. This could fundamentally change aspects of that species over thousands of years, from physiology to physical appearance. Also, the introduction of this new species that wasn't in the area previously could pose a threat to the organisms already there (e.g., disturb predator/prey balance);

- *Forests* will be negatively affected as warming temperatures change the frequency of wildfire and pestilent insects' outbreaks. The productivity of forests (i.e., the amount of oxygen they produce) will decrease due to changes in precipitation, temperature and the amount of carbon dioxide in the air;

- *Water resource* supply and quality will be altered by warming temperatures, changes in precipitation, and the rise in the sea level. This change in water resources can, in turn, affect other sectors such as energy production, human health, and agriculture;

- *Society* will be affected, especially in vulnerable places with high poverty and young, sick, or older adults. Cities are also sensitive to climate change because a high concentration of people in a very small space can be readily affected by these changes; and

- *International* impacts will be seen, as people in developing countries are more susceptible to climate change than those in already-developed countries. These changes in countries across the globe will no doubt affect their relationships with the United States regarding trade, tourism, and production of energy.

Reducing climate change

Reducing carbon dioxide emissions has become a great topic of discussion among both scientists and the general public. The Environmental Protection Agency has implemented various laws and regulations that help reduce carbon pollution emissions from power plants, automotive vehicles, and factories. In the 1990s, carbon dioxide emissions increased by about 1.3% each year, and they are expected to increase exponentially in the coming decades (an estimated 1.9-2.5% each year).

This increase in carbon dioxide emissions is due in part to the inability of *carbon sinks,* such as oceans, algae, and plants, to absorb the carbon dioxide as fast as it is being produced. Before reducing current carbon emissions can be tackled, plans to reduce the current increased rate of emissions must be conceived and implemented.

Automobiles currently produce around 30% of all carbon emissions, so changing the way transportation is used is a good place to start. Electric and hybrid cars recently came on the market to help decrease emissions by increasing the fuel efficiency (i.e., miles-per-gallon ratio) of automobiles.

Renewable energy sources such as hydroelectricity, wind power, and solar power have become dependable alternatives to fossil fuels. Other ideas for the reduction of global warming include funding programs and supporting laws and treaties that aid in reducing greenhouse gas emissions. Slowing the rate of deforestation is one way to ensure that excess carbon is naturally cycled and taken out of the atmosphere by plants.

Relevant laws and treaties

The National Oceanic and Atmospheric Administration (NOAA) is a scientific agency of the United States that focuses its efforts on understanding the condition of the atmosphere and the oceans. They research to warn the public of dangerous/extreme weather and guide the protection of coastal resources and the oceans. They also seek to improve understanding of the environment.

Along with the *Clean Air Act* that was passed in 1987, several other protocols, treaties, and panels exist that all act with the same concern: that anthropogenic forces may cause irreparable damage to the planet. These groups work to lessen carbon emissions, educate the public, and regulate industries.

- *The Clean Air Act,* a federal law that puts limits on the emissions of pollutants into the air. The pollution that is the biggest issue is *smog*, which poses a threat to human health and is created largely by emissions from automobiles. This legislation tightened the reins on many companies, which then had to comply with the new regulations or face stiff fines. The CAA has helped greatly in the reduction of air pollution in the United States;

- *The Climate Action Network* is a group of environmental organizations from around the globe working together to push countries to solve the issues that led to the onset of global climate change. They work together to strike international negotiations, pressure governments to work more quickly on passing their legislation and contribute to the development of the United Nations framework on climate change protocols;

- *The Montreal Protocol* was created in 1987 when all the major countries agreed to take measures to eliminate the production of ozone-depleting substances. This landmark global agreement to aid in the protection and restoration of the ozone layer has worked, and the levels of ODSs in the atmosphere are gradually decreasing; and

- *The Kyoto Protocol* was signed a decade after the Montreal Protocol and served the primary purpose of getting all industrialized countries (Europe, Canada, New Zealand, Japan, and several others) to reduce their emissions of carbon dioxide and other greenhouse gases.

Nearly all industrialized countries, except Australia and the United States, agreed to the treaty. The refusal to sign by the U.S. was due to fears that the protocol did not provide a significant plan to engage in larger developing countries, such as China, to commit to these standards.

Another possible reason is the concern that living up to the agreement could have a negative impact on the economy.

Notes

Loss of Biodiversity

By the year 2050, climate change alone is expected to threaten over one-quarter of land-dwelling species with extinction. Marine life and some freshwater species, especially those living in sensitive ecosystems (e.g., coral reefs), are also at risk for extinction due to elevated water temperatures. Over the past several millions of years, organisms have evolved and adapted to living in their environments.

When changes like temperature increase take place over a few short decades, evolved animals do not get enough time to adapt to these changes and may die out. Such events not only lead to the extinction of certain species but also threaten the existence of the other species in the ecosystem that rely on the now-extinct organisms as a source of food or as a higher predator in the food chain. This "domino effect" is often irreversible and highly destructive to whole ecosystems.

The *Intergovernmental Panel on Climate Change* (IPCC) has predicted that by the year 2100, the surface of the Earth will have warmed another 6 °C. This estimate is based on current trends of burning fossil fuels and toxic emissions. Though scientists cannot predict exactly how each species, and in turn, each ecosystem will respond, they do predict that the effects will be catastrophic.

In order to put an average global temperature change of 6 °C in perspective, think of the following: all of the changes that have occurred over the last few decades that have been ascribed to global warming — such as glacial melting, extreme storms, flooding, droughts, heatwaves, coral reef bleaching and dying, and so on — have occurred when the Earth's average global temperature changed by less than 1 °C. During the height of the last Ice Age roughly 20,000 years ago, the global temperature was only around 6 °C cooler than it is now.

Habitat loss

Each species on Earth has evolved over millions of years to best live and thrive in its ecological *niche,* an area with specific living conditions such as a range of temperatures and with suitable plants and animals for their consumption. Some species are better capable of adapting to sudden changes, though most are not.

For example, rats and cockroaches can live in a variety of conditions outside of their ideal niche, while pandas must live where there is bamboo, or they will not survive. Anthropogenic forces are no doubt causing rapid changes in global temperatures, disrupting the niches of many organisms, and threatening the lives of those unable to migrate quickly enough to a new suitable area. Several studies have indicated that a rise in the average global temperature of about 1.8-2 °C would result in the threat of extinction for over a million species. Unless greenhouse gas emissions are reduced drastically around the globe, habitat loss and the extinction of various organisms is inevitable.

Polar bears may become extinct due to the predicted disappearance of all Arctic sea ice over the 21st century. Polar bears are currently the world's largest predator on land. They can go months without eating because they build up enough fat to last through times when food is not readily available. Without sea ice, polar bears would not be the only species affected; much, if not all, of the Arctic ecosystem, would collapse.

Figure 7.10. *A polar bear family is living with the effects of climate change on habitat loss. Normally, this area would be covered with glacial ice.*

Coral reefs are affected by coral bleaching, a phenomenon that occurs when rising temperatures or ocean acidity stresses a coral. Under normal conditions, coral contains microscopic algae called zooxanthellae that provide the reef with a source of food and give the coral its vibrant colors.

When stressed, the coral will expel the zooxanthellae and turn white in color. If the zooxanthellae do not return to that coral, it will not receive food and eventually die. Changes in water temperature as small as 1 °C can stress a coral to the point of bleaching.

In 2002, approximately 60% of the Great Barrier Reef in Australia was affected by bleaching. Unless temperatures stop rising, much of this beautiful reef will be dead in a matter of decades, as well as the hundreds of species who live on the reef and use it as a source of food and shelter.

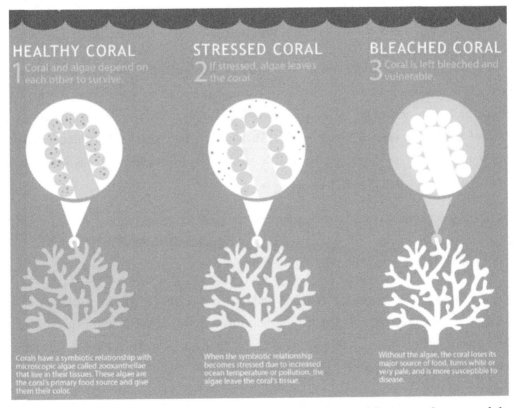

Figure 7.11. *An overview of the process of coral bleaching. If the stress that caused the bleaching is not severe, the coral may recover.*

Plants, like all insects and animals, require a specific climate to grow and thrive. For example, a silver maple will not be found growing next to a Saguaro Cactus; they are two very different species that have very different physiological needs.

Changes in temperature and precipitation will affect the ability of some species to continue to thrive. Unlike animals and insects, plants cannot move quickly to migrate to a more suitable niche.

Therefore, climate change may occur at a rate that is too rapid to allow for adaptation by most plants. The loss of these plants will affect the food chain greatly, as some other organisms very likely relied on that plant as a source of nutrition and/or shelter.

Overuse

A loss of biodiversity can occur through the overexploitation or overuse of species. One such example is the overuse of marine species like invertebrates and fish. Overexploitation, including both over-harvesting and overfishing, have depleted nearly 75% of all marine species in the world. Anchovy off the coast of Peru and cod from Newfoundland are no longer viable food resources because their stock has been almost entirely depleted.

As populations grow and the demand for food increases, humans have moved to other sources of marine life to meet their nutritional needs. Some overexploitation of marine animals is wasteful, like the finning industry common to parts of Asia.

Dolphins and sharks are captured, and their fins removed to make a delicacy called fin soup. The animals are either killed for their fins or released back into the ocean, where they often die from blood loss and the inability to swim properly. Finning has gained national attention in the past few years, and organizations have moved to put an end to this practice.

Pollution

Although the effects of pollution on animals have not been as widely tested and measured as its effects on humans, it is safe to assume that animal life is affected by pollution just as much as humankind; holes in the ozone layer can allow harmful UV rays to reach the Earth's surface, causing skin cancer in people, as well as damage to crops and other organisms.

One major phenomenon that affects biodiversity is *acid rain,* rain containing sulfuric and nitric acid caused by pollution from various sources that burn fossil fuels and release emissions into the air. This acidic water vapor travels into the atmosphere, where it eventually forms clouds and returns to Earth as acidic rainfall.

Acid rain has a detrimental effect on streams and lakes, as the organisms that reside there are not suited to live in acidic environments. It can also harm plant life by damaging the leaves and branches of trees, slowing their growth.

Introduced species

An *introduced species,* also called an exotic species, is a plant or animal that is not native to the area where it is introduced. These new species can arrive at the new locality either through deliberate action or through accidental transportation (e.g., North American tumbleweed).

Non-native species can affect the local ecosystem in various ways; some may have little or no impact; some have a negative impact and others a positive impact. Some have been introduced to ecosystems on purpose as a method of biological pest control and are used as an alternative to chemical pesticides on crops.

One example of an introduced species is the alfalfa weevil, a serious foraging pest that was substantially reduced through the introduction of its natural enemies.

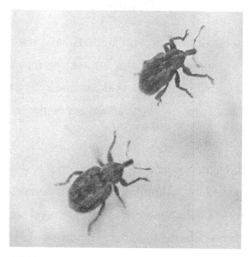

Figure 7.12. *Adult alfalfa weevils threatened the native species through foraging.*

Introduced species that become harmful or are considered a pest in the new area are termed *invasive species* and can cause great harm to an ecosystem. They can grow quickly and overcrowd areas in a matter of days or weeks. Scientists are concerned that invasive species will begin to take over areas and force other native plants to extinction.

For example, Japanese stilt grass is an invasive species that was unintentionally introduced to the United States when used as a packaging material for shipments from China. There is fear that the unstoppable growth of this grass, which can grow in a variety of habitats, will crowd out many native species.

The Asian long-horned beetle was introduced to the U.S. unintentionally via cargo from Asia to New York. It is destructive to maple and other hardwoods. Another example is Burmese pythons, which were kept by people as pets and later got released or escaped into the wilderness of South Florida. They spread quickly and now threaten the existence of many native species (such as the Key Largo woodrat, deer, and indigo snakes) and disrupt the indigenous food chain balance.

Endangered and extinct species

An *endangered species* (see table below) is a plant or animal that is at great risk for extinction. They are organisms that have been "red-listed," meaning that efforts to conserve their species are of the utmost importance.

An endangered species is the second most severe status for an organism, after those that have been deemed *critically endangered* (those that face the highest risk of extinction in the wild). The number of species on the list changes from time to time; some may be taken off if efforts have been successful in conserving their kind, and they are no longer in danger of extinction.

Group	United States			Foreign			Total Listings (U.S. + Foreign)	U.S. Listings with active recovery plans
	Endangered	Threatened	Total Listings	Endangered	Threatened	Total Listings		
Annelid worms	0	0	0	0	0	0	0	0
Flatworms and roundworms	0	0	0	0	0	0	0	0
Sponges	0	0	0	0	0	0	0	0
Corals	0	6	6	0	16	16	22	0
Snails	38	12	50	1	0	1	51	29
Amphibians	20	15	35	8	1	9	44	21
Reptiles	15	24	39	69	20	89	128	36
Mammals	75	25	100	254	20	274	374	64
Clams	75	13	88	2	0	2	90	71
Fishes	93	70	163	19	3	22	185	105
Arachnids	12	0	12	0	0	0	25	12
Insects	64	11	75	4	0	4	79	41
Crustaceans	22	3	25	0	0	0	25	18
Birds	79	21	100	217	17	234	334	86
Millipedes	0	0	0	0	0	0	0	0
Hydroids	0	0	0	0	0	0	0	0
Animal Totals	493	200	693	574	77	651	1344	483
Conifers & cycads	2	2	4	0	2	2	6	3
Flowering plants	700	161	861	1	0	1	862	646
Ferns and allies	29	2	31	0	0	0	31	26
Lichens	2	0	2	0	0	0	2	2
Plant Totals	733	165	898	1	2	3	901	677
Grand Totals	1226	365	1591	575	79	654	2245	1160

Figure 7.13. *Table showing organisms around the globe and the number of species that are considered endangered or threatened. Over 1226 plants and animals in the United States were considered endangered as of November 28, 2015.*

Despite the best efforts of individuals and organizations across the globe, some species still go extinct. *The Natural Extinction Rate* is the rate at which species would become extinct if humans were out of the picture. Scientists estimate that the rapid loss of species today is somewhere between 1,000 and 10,000 times higher than the natural extinction rate.

Experts have calculated that between 0.01 and 0.1% of species will become extinct this year. If estimates of the total number of living species are correct (because we are discovering new species all the time), between 10,000 and 100,000 species will become extinct in the next year.

Mass extinction events occur from time to time, involving a catastrophic disaster that wipes out several species at once. An example is the cataclysmic impact of the huge meteor striking the Earth that is believed to have killed off species of dinosaurs around 65 million years ago.

Maintenance through conservation

The biodiversity that we see on planet Earth today is the result of billions of years of evolution. Unfortunately, humanity has over-exploited natural resources, resulting in disturbances in the natural environment. Scientists estimate that the extinction rate of species is between 1,000 and 10,000 times higher than it would be without any human interference.

Conserving biodiversity is about ensuring that natural ecosystems are maintained and that the populations, species, genes, and the array of interactions between them survive and persist into the future. The conservation of biodiversity relies heavily on several disciplines of the natural sciences and biological sciences, as well as law, public policy, economics, and other disciplines working together. Conservation biology has grown as a scientific discipline in the past several decades and has helped to broaden our understanding of biodiversity and natural ecosystems.

One method of preserving biodiversity is the conservation of rainforests. Rainforests are the biggest stores of the planet's biodiversity and act as wonderful climate change regulators. They filter pollution by absorbing very large amounts of carbon dioxide from the atmosphere. Rainforests are rich with plant life, which require CO_2 for the process of photosynthesis. By removing excess carbon dioxide from the atmosphere, rainforests help keep the climate stable and functioning properly.

Other methods of biodiversity conservation include incentives by governments to private landowners to practice conservation, as well as management practices on the part of local and state parks, sanctuaries, and wildlife services. These incentives may be monetary and offered as a part of the government's environmental protection programs.

In some parts of the world, countries have begun to construct *wildlife corridors.* These look like large bridges that cross over highways and other routes of transportation. Before the construction of these bridges, wildlife could only reach the other side if they tried to cross the road itself, putting themselves in the way of danger and possibly death. This not only threatens the animals but the lives of the motorists on the road.

Every year, more than 200 people are killed and thousands injured in animal collisions. Insurance companies estimate that the annual cost to society for these accidents in deaths and injuries is over $200 million.

Figure 7.14. *A wildlife crossing over a busy highway that allows for the safe crossing of deer, bears, reptiles, and other animals (left); an underground tunnel crossing for tortoises (right).*

Seed banks are a method of plant conservation that entails storing seeds in cold and dry conditions. This allows the seeds to remain dormant while prolonging their viability. Seeds are collected from thousands of plants and are stored for future use, including wild, ornamental, medicinal, and crop-based seeds.

The Millennium Seed Bank Project at the Royal Botanical Gardens in Great Britain stores nearly 10% of the world's plant diversity. Seed banks provide quick access to plant samples, which allow scientists to evaluate them on nutritional, medicinal, and genetic levels. In a seed bank, they are almost entirely immune to disease, predators, and other factors that endanger their species.

A recent example of a seed bank can be found in Svalbard, Norway, where there is a project to house seeds from all known plants to preserve plant biodiversity in the event of a global cataclysm.

In the more recent past, *cryopreservation* has become a more widely used process, and it involves the storage of living tissue at extremely low temperatures (around −196 °C). This procedure is available for more than 150 different species of plants, though for each species, the method must be fine-tuned and empirically adapted. When stored properly, seeds may remain viable for hundreds or even thousands of years.

The steps of cryopreservation include the use of young zygotic embryos, the addition of retardants or inhibitors that slow down cell growth, an overlay of mineral oil after placement in a low-temperature storage facility, and the reduction of oxygen tension.

Cryopreservation has many uses, some of which include:

- Conservation of endangered plant life through storage of seeds in the seed bank;

- Plant pathogen conservation;

- Storing genetically altered tissues;

- Preservation and storage of plant tissue that has medicinal value;

- Conservation of plant germplasm; and

- Storage for use in genetic studies.

Relevant laws and treaties

The *Endangered Species Act* (ESA) of 1973 serves the purpose of conserving the species around the world that are threatened or endangered, as well as working to conserve the ecosystems they depend on to live. In 1972, President Nixon stated that the United States did not have suitable conservation efforts and needed to aim toward preventing the extinction of species.

Congress responded and passed the Endangered Species Act (ESA). It recognized that "*our rich natural heritage is of esthetic, ecological, educational, recreational and scientific value to our Nation and its people*" and expressed great concern that the native plants and animals of the United States were in danger of going extinct. Through the ESA, species may be listed as either threatened or endangered. All species of animals and plants are eligible to be listed if need be.

The Marine Mammal Protection Act (MMPA) was enacted in 1972, and it extends protection to all marine mammals. With certain exceptions, the MMPA prohibits both the taking of marine mammals out of U.S. waters, as well as the importation of marine mammals or products into the United States. Congress passed the Act by a few principles:

- That some marine mammals may be endangered as a result of human activity;

- Marine mammals are grand resources for all nations;

- These species must not be depleted or allowed to fall below their optimum sustainable population level; and

- Steps should be taken to replenish the species that are endangered.

The World Wildlife Foundation (WWF) works to advance policies to aid in the fight against climate change, to help nature and people adapt to a changing climate, and to engage with businesses and industries to reduce carbon emissions.

The WWF also has several other programs in place, such as Adopt a Polar Bear, Adopt a Tiger, Adopt an Elephant, and several other "adoption" choices of animal species that are at risk due to climate change.

Notes

Appendix

Glossary of Terms

A

Abated – reduced by a degree or in intensity; elimination of pollution.

Abiotic – non-living chemical and physical factors of the environment; see *Biotic*.

Abiotic factor – an environmental factor that is nonliving, such as water, soil, temperature or sunlight.

Absorption – one substance is taking in another, either physically or chemically.

Absorption pit (soakaway) – a hole dug in the permeable ground and filled with broken stones or granular material and usually covered with earth, allowing the collected water to soak into the ground.

Abundance – the number or amount of something.

Acclimation – the process of an organism adjusting to a change in its environment.

Acids – substances that release hydrogen ions (H^+ or protons) in water.

Acid mine drainage – the outflow of acidic water from metal mines or coal mines.

Acid precipitation – acidic rain, snow, or dry particles deposited from the air due to increased acids released by anthropogenic or natural resources; see *Acid rain*.

Acid rain – rain (and snow, fog, dust particles, etc.) containing acids that form in the atmosphere when sulfur dioxides and nitrogen oxides from industrial emissions and automobile exhaust combine with water; see *Acid precipitation*.

Active solar system – a mechanical system that actively collects, concentrates, and stores solar energy.

Acute/Chronic – acute is a short, one-time exposure; chronic is a continuous, low-level exposure.

Acute poverty – insufficient income or access to resources needed to provide the necessities for life such as food, shelter, sanitation, clean water, medical care, and education.

Adapted – to become accustomed to the natural factors that are in each area and to be able to survive these factors; an organism can be either positively or negatively adapted.

Adaptation – a characteristic of an organism that has been favored by natural selection.

Adaptive management – a management plan designed from the outset to "learn by doing," and to actively test hypotheses and adjust treatments as new information becomes available.

Adaptive radiation – closely related species that look very different as a result of having adapted to widely different ecological niches.

Administrative courts – courts that hear enforcement cases for agencies or consider appeals to agency rules.

Administrative law – executive orders, administrative rules and regulations and enforcement decisions by administrative agencies and special administrative courts.

Adsorption – one substance taking up another at its surface.

Aerobic – requiring air or oxygen; used in reference to decomposition processes that occur in the presence of oxygen.

Aerosol – a suspension of solid or liquid particles within the air; some human-made aerosols (dust particles) in the atmosphere are believed to reflect solar radiation, therefore producing a cooling effect on global temperatures, while others can also absorb solar radiation (depending on the chemical's properties) to create a warming effect.

Aesthetic degradation – changes in environmental quality that offend our aesthetic senses.

Affluenza – as defined in the book of the same name: 1) the bloated, sluggish and unfulfilled feeling that results from efforts to keep up with the Joneses; 2) an epidemic of stress, overwork, waste and indebtedness caused by dogged pursuit of the Australian dream; 3) an unsustainable addiction to economic growth; the general Western environmentally unfriendly high-consumption lifestyle; a combination of words "affluence" and "influenza."

Afforestation – planting new forests on lands that have not been recently forested.

Agroforestry – (sustainability) an ecologically based farming system that, through the integration of trees onto farms, increases social, environmental and economic benefits to land users.

Air pollution – the modification of the natural characteristics of the atmosphere by a chemical, particulate matter or biological agent.

Alaska Pipeline – built from April 29, 1974, to June 20, 1977, this above-ground pipeline through Alaska brings oil from the oil wells in northern Alaska to shipping ports in southern Alaska.

Albedo – reflectance; the ratio of light from the Sun that is reflected by the Earth's surface to the light received by it; unreflected light is converted to infrared radiation (heat), which causes atmospheric warming; surfaces with a high albedo, like snow and ice, generally contribute to cooling, whereas surfaces with a low albedo, like forests, generally contribute to warming; changes in land use that significantly alter the characteristics of land surfaces can alter the albedo; see *Radiative forcing*.

Algae – any of the various chiefly aquatic, eukaryotic, photosynthetic organisms, ranging in size from single-celled forms to the giant kelp; once considered to be plants but now classified separately because they lack true roots, stems, leaves, and embryos.

Algal bloom – the rapid and excessive growth of algae; generally caused by high nutrient levels combined with other favorable conditions; blooms can deoxygenate the water, leading to the loss of wildlife.

Alien species – any species of plant or animal that is living outside of its natural distribution; see *Introduced species*.

Alloy – composite blend of materials made under special conditions; metal alloys like brass and bronze are well known, but there are also many plastic alloys.

Alternative fuels – fuels like ethanol and compressed natural gas that produce fewer emissions than traditional fossil fuels.

Alternative renewable energy – renewable energy sources are those that are replenished by natural processes at rates exceeding their rate of use for human purposes, unlike *Fossil fuels* which are not replenished at a useful rate; include solar, biomass, wind, geothermal and hydropower.

Ambient air – the air immediately surrounding a location.

Amino acid – units or building blocks that make peptide and protein molecules; an organic compound containing an amino group and a carboxyl group.

Ammonia – (NH_3) a pungent, colorless, gaseous alkaline compound of nitrogen (N) and hydrogen (H) that is very soluble in water and can easily be condensed to a liquid by cold and pressure.

Anaerobic – not requiring air or oxygen; used in reference to decomposition processes that occur in the absence of oxygen; see *Anaerobic digestion* and *Anaerobic respiration*.

Anaerobic digestion – the biological degradation of organic materials in the absence of oxygen to yield methane gas (that may be combusted to produce energy) and stabilized organic residues (that may be used as a soil additive).

Anaerobic respiration – the incomplete intracellular breakdown of sugar or other organic compounds in the absence of oxygen that releases some energy and produces organic acids or alcohol.

Ancient forest – a forest that has been free from disturbances long enough to have the characteristics of equilibrium ecosystems, including mature trees and species diversity; see *Old-growth forest*.

Anemia – low levels of hemoglobin due to an iron deficiency or lack of red blood cells.

Annual – a plant that lives for a single growing season.

Anomalies – deviation from the normal order or general rule.

Anoxic – with abnormally low levels of oxygen.

Anthropocentric – 1) the belief that humans hold a special place in nature; 2) being centered primarily on humans and human affairs.

Anthropogenic – originating in or due to human activity (i.e., human-made, not natural).

Anthroposophy – spiritual philosophy based on the teachings of Rudolf Steiner (1861 – 1925) which postulates the existence of an objective, intellectually comprehensible spiritual world accessible to direct experience through inner development or, more specifically, through conscientiously cultivating a form of thinking independent of sensory experience; Steiner was the initiator of "biodynamic gardening," which is similar to organic gardening with the added belief that animals, crops, and soil are essentially one organism.

Application efficiency – (sustainability) the efficiency of watering after losses due to runoff, leaching, evaporation, wind, etc.

Appropriate technology – technology that can be made at an affordable price by ordinary people using local materials to do useful work in ways that do the least possible harm to both human society and the environment.

Appropriated carrying capacity – another name for the Ecological Footprint, but often used in referring to the imported ecological capacity of goods from overseas.

Aquaculture – the cultivation of aquatic organisms under controlled conditions.

Aqueduct – a pipe or channel designed to transport water from a remote source, usually by gravity; a bridge-like structure supporting a conduit or canal passing over a river or low ground.

Aquifer – a bed or layer yielding water for wells and springs etc.; an underground geological formation capable of receiving, storing and transmitting large quantities of water; types include: confined (sealed and possibly containing "fossil" water), unconfined (capable of receiving inflow) and Artesian (an aquifer in which hydraulic pressure causes the water to rise above the upper confining layer).

Arable land – land that can be used for growing crops.

Arbitration – a formal process of dispute resolution in which there are stringent rules of evidence, cross-examination of witnesses, and a legally-binding decision made by the arbitrator that all parties must obey.

Arithmetical growth – a pattern of growth that increases at a constant amount per unit time (e.g., 1, 2, 3, 4 or 1, 3, 5, 7).

Artesian well – the result of a pressurized aquifer intersecting the surface or being penetrated by a pipe or conduit, from which water gushes without being pumped; also called a "spring."

Asbestos – a fibrous incombustible mineral known to cause fibrosis and scarring in the lungs; a known carcinogenic material that causes lung cancer and mesothelioma.

Ash – the grayish-white to black powdery residue left when something is burned.

Asthma – a distressing disease characterized by shortness of breath, wheezing and bronchial muscle spasms.

Aswan High Dam – dam across the Nile River in Egypt, which impounds one of the largest reservoirs in the world; the artificial lake created by the dam called Lake Nasser inundated many villages along the Nile and submerged some of the pyramids; hydroelectric installations were added in 1960 to the Aswan Dam.

Atmosphere – general name for the layer of gases around a material body; the Earth's atmosphere consists, from the ground up, of the troposphere (which includes the planetary boundary layer or peplosphere - the lowest layer), the stratosphere, the mesosphere, the ionosphere (or thermosphere), the exosphere and the magnetosphere.

Atmospheric deposition – sedimentation of solids, liquids or gaseous materials from the air.

Atom – the smallest unit of matter that has the characteristics of an element; consists of three main types of subatomic particles: protons, neutrons, and electrons.

Atomic number – the characteristic number of protons per atom of an element; used as an identifying attribute.

Auto emissions standards – the standards that are set to regulate how much pollution by one's vehicle is permissible.

Autotroph – an organism that produces complex organic compounds from simple inorganic molecules using energy from light or inorganic chemical reactions.

Available energy – energy with the potential to do work; also called "exergy."

Available water capacity – that proportion of soil water that can be readily absorbed by plant roots.

Avoidance – (sustainability) the first step in the waste hierarchy where waste generation is prevented (avoided).

B

"Baby Boom" – a sudden large increase in the birthrate over a period, specifically the 15 years after World War II.

Backflow – the movement of water back to its source (e.g., contaminated water in a plumbing system).

Bacteria – any of a group (as kingdom *Prokaryotae*) of prokaryotic unicellular round, spiral or rod-shaped single-celled microorganisms that are often aggregated into colonies or motile by means of flagella; live in soil, water, organic matter or the bodies of plants and animals and are autotrophic, saprophytic or parasitic in nutrition; important because of their biochemical effects and pathogenicity.

Baffle – (landscape design) an obstruction to trap debris in drainage water.

Ballast – anything that serves no particular purpose except to give bulk or weight to something or that provides additional stability.

Bagasse – the fibrous residue of sugarcane milling, used as a fuel to produce steam in sugar mills.

Barometric pressure – atmospheric pressure as indicated by a barometer.

Barrier islands – low, narrow, sandy islands that form offshore from a coastline.

Baseload – the steady and reliable supply of energy through the grid; punctuated by bursts of higher demand known as "peak-load;" supply companies must be able to respond instantly to extreme variation in demand and supply, especially during extreme conditions; gas generators can react quickly while coal is slow but provides a steady "baseload;" *Renewable energies* are generally not available on-demand in this way.

Bases – substances that bond readily with hydrogen ions.

Basin – a large, bowl-shaped depression on the surface of the land or on the ocean floor.

BAT – acronym for "*Best available, economically achievable technology.*"

Batesian mimicry – evolution by one species to resemble the coloration, body shape or behavior of another species that is protected from predators by a venomous stinger, being inedible or some other defensive adaptation.

Batters – (landscape design) the slope of earthworks such as drainage channels.

Benthos – the bottom of a sea or lake.

Best available, economically achievable technology (BAT) – the least pollution-causing technological solution; also known as "best available techniques," "best practicable means" or "best practicable environmental option."

Best practical control technology (BPT) – the best technology for pollution control available at a reasonable cost and operable under normal conditions.

Best practice – a process, technique, or innovative use of technology, equipment, resources, or other measurable factors that have a proven record of success.

Beta particles – high-energy electrons released by radioactive decay.

Bhopal, India – a noxious gas (methylisocyanate) blanketed the city when the water got into a tank containing 40 tons of MIC, setting off a chemical reaction; 1,754 died, with over 200,000 injured.

Bill – a piece of legislation introduced in Congress and intended to become law.

Bioaccumulation – the accumulation of a substance, such as a toxic chemical, in the tissues of a living organism in higher concentrations in the organism than in its direct environment or food.

Biocapacity – a measure of the biological productivity of an area; it may depend on natural conditions or human inputs like farming and forestry practices; the area needed to support the consumption of a defined population.

Biocoenosis (alternatively, biocoenose or biocenose) – all the interacting organisms living together in a specific habitat (or biotope).

Biocentric preservation – a philosophy that emphasizes the fundamental right of living organisms to exist and to pursue their goods.

Biocentrism – the belief that all creatures have rights and values; being centered on nature rather than humans.

Biocide – a broad-spectrum poison that can kill a wide range of organisms.

Biodegradable – capable of being decomposed through the action of organisms, especially bacteria.

Biodegradable plastics – plastics that can be decomposed by microorganisms.

Biodiversity – the variety of life in all its forms, levels, and combinations; includes ecosystem diversity, species diversity, and genetic diversity.

Bioelement – an element that is found within a living organism.

Bioenergy – in its most narrow sense, a synonym for biofuel (i.e., fuel derived from biological sources); in its broader sense, also encompasses biomass, the biological material used as a biofuel, as well as the social, economic, scientific and technical fields associated with using biological sources for energy.

Biofuel – the fuel produced by the chemical and/or biological processing of biomass; can be a solid (e.g., charcoal), a liquid (e.g., ethanol) or a gas (e.g., methane).

Biogas – landfill gas and sewage gas; also called "biomass gas."

Biogeochemical cycle – 1) a circuit or pathway by which a chemical element or molecule moves through both biotic ("bio-") and abiotic ("geo-") parts of an ecosystem; 2) the movement of matter within or between ecosystems; caused by living organisms, geological forces or chemical reactions (e.g., the cycling of nitrogen, carbon, sulfur, oxygen, phosphorus and water).

Biogeographical area – an entire self-contained natural ecosystem and its associated land, water, air, and wildlife resources.

Biological community – the populations of plants, animals, and microorganisms living and interacting in a certain area at a given time.

Biological controls – use of natural predators, pathogens or competitors to regulate pest populations.

Biological or biotic factors – organisms and products of organisms that are part of the environment and potentially affect the lives of other organisms.

Biological oxygen demand (BOD) – a standard test for measuring the amount of dissolved oxygen utilized by aquatic microorganisms.

Biological pest control – a method of controlling pests (including insects, mites, weeds, and plant diseases) that relies on predation, parasitism, herbivory or other natural mechanisms.

Biological pests – organisms that reduce the availability, quality or value of resources useful to humans.

Biological productivity – (bioproductivity) the capacity of a given area to produce biomass; different ecosystems (e.g., pasture, forest, etc.) will have different levels of bioproductivity; determined by dividing the total biological production by the total area available.

Biological resources – the plants and animals that are important for the various services they provide.

Biologically productive land – land that is fertile enough to support forests, agriculture and/or animal life; all the biologically productive land of a country comprises its biological capacity; arable land is typically the most productive area.

Biomagnification – an increase in the concentration of certain stable chemicals (e.g., heavy metals or fat-soluble pesticides) in successively higher trophic levels of a food chain or web.

Biomass – 1) the materials derived from photosynthesis (fossilized materials may or may not be included) such as forest, crops, wood and wood wastes, animal wastes, livestock operation residues, aquatic plants, and municipal and industrial wastes; 2) the quantity of organic material present in a unit area at a particular time, mostly expressed as tons of dry matter per unit area; 3) organic matter that can be used as fuel.

Biomass fuel – organic material produced by plants, animals or microorganisms that can be burned directly as a heat source or converted into a gaseous or liquid fuel.

Biomass pyramid – a metaphor or diagram that explains the relationship between the amounts of biomass at different trophic levels.

Biome – a climatically and geographically defined area of ecologically similar communities of plants, animals, and soil organisms often referred to as ecosystems.

Biophysical – the living and non-living components and processes of the ecosphere; biophysical measurements of nature quantify the ecosphere in physical units such as cubic meters, kilograms, or joules.

Bioregion – an area comprising a natural ecological community and bounded by natural borders; see *Ecoregion*.

Bioremediation – a process using organisms to remove or neutralize contaminants (e.g., gasoline), mostly in soil or water.

Biosolids – nutrient-rich organic materials derived from wastewater solids (sewage sludge) that have been stabilized through processing.

Biosphere – 1) the zone of air, land, and water at the surface of the Earth that is occupied by living organisms; 2) the combination of all ecosystems on Earth maintained by the energy of the Sun; 3) the interface between the hydrosphere, the geosphere and the atmosphere.

Biosphere reserves – world heritage sites identified by the *International Union for the Conservation of Nature* (IUCN) as worthy for national park or wildlife refuge status because of high biological diversity or unique ecological features.

Biota – all organisms in a given area.

Biotic – relating to, produced by or caused by living organisms; compare with *Abiotic*.

Biotic potential – the maximum reproductive rate of an organism, given unlimited resources and ideal environmental conditions; compare with *Environmental resistance*.

Birth control – 1) any method used to reduce births, including celibacy, delayed marriage, and contraception; 2) devices or medication intended to prevent the implantation of fertilized zygotes; includes induced abortions.

Birth rate – the number of people born as a percentage of the total population in any given period; expressed in the number of live births per 1,000 people.

Black lung disease – inflammation and fibrosis caused by the accumulation of coal dust in the lungs or airways.

Blackwater – household wastewater that contains solid waste (i.e., toilet discharge) that cannot be reused without purification.

Blue revolution – new techniques of fish farming that may contribute as much to human nutrition as miracle cereal grains but also may create social and environmental problems.

Bluewater – collectible water from rainfall; the water that falls on roofs and hard surfaces, usually flowing into rivers and the sea and recharging the groundwater; in nature the global average proportion of total rainfall that is blue water is about 40%; bluewater productivity in the garden can be increased by improved irrigation techniques, soil water storage, moderating the climate, using garden design and water-conserving plantings; also safe use of *Greywater*.

Bog – an area of waterlogged soil that tends to be peaty; fed mainly by precipitation; low productivity; some bogs are acidic.

Boreal – 1) of the north or northern areas; 2) referring to the cold temperate Northern Hemisphere forests of birch, poplar, or conifers that grow where there is a mean annual temperature < 0 °C.

Boreal forest –broadband of mixed coniferous and deciduous trees that stretches across northern North America (and Europe and Asia); its northernmost edge, the taiga, intergrades with the arctic tundra.

BPT – an acronym for "*Best practicable control technology;*" the best technology available for pollution control at a reasonable cost and operable under normal conditions.

Brackish water – fresh and saltwater combined.

Breeder reactor – 1) a nuclear reactor that produces more fuel than it consumes; this kind of reactor is used mainly to produce plutonium; 2) a nuclear reactor that produces fuel by bombarding isotopes of uranium and thorium with high-energy neutrons that convert inert atoms to fissionable ones.

Breeding – a group of organisms having common ancestors and certain distinguishable characteristics, especially a group within a species developed by artificial selection and maintained by controlled propagation.

Broad-acre farm – commercial farm covering a large area; usually a mixed farm in dryland conditions.

Brownfields – abandoned or underused urban areas in which redevelopment is blocked by liability or financing issues related to toxic contamination.

Brundtland Commission Report – a UN report, *Our Common Future*, published in 1987 and dealing with sustainable development and the policies required to achieve it, which the report characterizes as "development that meets the needs of the present without compromising the ability of future generations to meet their own needs."

C

C3 plants – about 95% of all plants; form molecules with three carbon items during photosynthesis; as CO_2 levels increase, C3 plants increase their photosynthesis; compare with *C4 plants*.

C4 plants – comprise about 5% of all plants; during photosynthesis, they form molecules with four carbon atoms and saturate at a given CO_2 level; most abundant in hot and arid conditions and include crops like sugar cane and soybeans; compare with *C3 plants*.

Calorie – a basic measure of energy that has been replaced by the SI unit the *Joule*; in physics it approximates the energy needed to increase the temperature of 1 gram of water by 1 °C which is about 4.184 joules; the Calories in food ratings (spelled with a capital C), and nutrition is 'big C' Calories or kcal.

Calorific value – the energy content of a fuel measured as the heat released on complete combustion.

Cancer – a group of diseases in which cells are aggressive (grow and divide without respect to normal limits), invasive (invade and destroy adjacent tissues) and sometimes metastatic (spread to other locations in the body).

Caprock – last layer of material on top of a geological formation such as the Canadian Shield.

Capillary action (wicking) – water is drawn through a medium by surface tension.

Capillary water – water that clings in small pores, cracks and spaces against the pull of gravity (e.g., the water retained in a sponge).

Capital – any form of wealth, resources, or knowledge available for use in the production of more wealth.

Captive breeding – raising plants or animals in zoos or other controlled conditions to produce stock for subsequent release into the wild.

Carbohydrate – an organic compound consisting of a ring or chain of carbon atoms with hydrogen and oxygen attached (e.g., sugars, starches, cellulose, and glycogen).

Carbon budget – a measure of carbon inputs and outputs for a particular activity.

Carbon credit – a market-driven way of reducing the impact of greenhouse gas emissions; allows an agent to benefit financially from an emission reduction; there are two forms of carbon credit, those that are part of national and international trade and those that are purchased by individuals; internationally, to achieve *Kyoto Protocol* objectives, "caps" (limits) on participating countries' emissions were established; to meet these limits the countries, in turn, set caps (allowances or credits: 1 convertible and transferable credit = 1 metric ton of CO_2-e emissions) for operators; operators that meet the agreed caps can then sell their unused credits to those operators who exceed the caps; operators can then choose the most cost-effective way of reducing emissions; individual carbon credits would operate in a similar way.

Carbon cycle – the *Biogeochemical cycle* by which carbon is exchanged between the biosphere, the geosphere, the hydrosphere and the atmosphere of the Earth; includes photosynthesis, decomposition and respiration (e.g., carbon dioxide is taken from the atmosphere by photosynthesizing plants and returned by the respiration of plants and animals and by the combustion of fossil fuels).

Carbon dioxide – a gas with the chemical formula CO_2 (the most abundant greenhouse gas emitted by fossil fuels).

Carbon Dioxide Equivalent (CO₂e) – the unit used to measure the impacts of releasing (or avoiding the release of) the seven different greenhouse gases; obtained by multiplying the mass of the greenhouse gas by its global warming potential (e.g., this would be 21 for methane and 310 for nitrous oxide).

Carbon equivalent (C-e) – obtained by multiplying the CO_2-e by the factor 12/44.

Carbon footprint – a measure of the carbon emissions that are emitted over the full lifecycle of a product or service, usually expressed as grams of CO_2-e.

Carbon labeling – use of product labels that display greenhouse emissions associated with the product.

Carbon management – storing CO_2 or using it in ways that prevent its release into the air.

Carbon monoxide (CO) – colorless, odorless, nonirritating but highly toxic gas produced by incomplete combustion of fuel, incineration of biomass or solid waste or partially anaerobic decomposition of organic material.

Carbon neutral – activities whose net carbon inputs and outputs are the same (e.g., assuming a constant amount of vegetation on the planet, burning wood will add carbon to the atmosphere in the short term, but this carbon will later cycle back into new plant growth).

Carbon pool – a storage reservoir of carbon.

Carbon sinks – places of carbon accumulation, such as large forests (organic compounds) or ocean sediments (calcium carbonate); carbon is removed from the carbon cycle for moderately long to very long periods; opposite of a *Carbon source*.

Carbon source – the originating point of carbon that reenters the carbon cycle; cellular respiration and combustion; opposite of a *Carbon sink*.

Carbon stocks – the quantity of carbon held within a carbon pool at a specified time.

Carbon taxes – a surcharge on fossil fuels that aims to reduce carbon dioxide emissions.

Carcinogen – a substance, radionuclide or radiation that is an agent directly involved in the promotion of cancer or the facilitation of its propagation.

Carnivores – an organism that eats only or primarily the meat of other organisms.

Carpooling – when people ride in cars together to help reduce emissions and traffic.

Carrying capacity – the maximum number of individuals of any species that can be supported by an ecosystem on a long-term basis.

Case law – precedents from both civil and criminal court cases.

Cash crops – crops that are sold rather than consumed or bartered.

Casks – barrels that are used to store spent fuels.

Catalytic converter – a reaction chamber typically containing a finely-divided platinum-iridium catalyst into which exhaust gases from an automotive engine are passed together with excess air so that carbon monoxide and hydrocarbon pollutants are oxidized to carbon dioxide and water.

Catastrophic systems – 1) dynamic systems that jump abruptly from one seemingly steady state to another without any immediate changes; 2) the detrimental effect that something, perhaps a natural disaster, has on an environment, destroying the ecosystem and surrounding living conditions.

Catchment area – the area that is the source of water for water supply, whether a dam or rainwater tank.

Cell – (biology) the structural and functional unit of all known living organisms and is the smallest unit of an organism that is classified as living

Cellular respiration – the process by which a cell breaks down sugar or other organic compounds to release energy used for cellular work; may be anaerobic or aerobic, depending on the availability of oxygen.

CERCLA (Superfund) Act of 1980 – sets up a fund to clean up abandoned hazardous waste sites; establishes strict liability, which means that any individual or corporation associated with the site can be held liable for the entire cost of the cleanup, regardless of their contribution to the pollution at the site; sets guidelines on how to clean up sites.

CFC (Chlorofluorocarbons) – a series of hydrocarbons containing both chlorine and fluorine; have been used as refrigerants, blowing agents, cleaning fluids, solvents and fire extinguishing agents; have been shown to cause stratospheric ozone depletion and have been banned for many uses; potent greenhouse gases which are not regulated by the *Kyoto Protocol* since the *Montreal Protocol* covers them.

Chain reaction – a self-sustaining reaction in which the fission of nuclei produces subatomic particles that cause the fission of other nuclei.

Channelization – to straighten using a channel.

Chaotic systems – systems that exhibit variability, which may not necessarily be "random," but whose complex patterns are not discernible over a normal human timescale.

Chemical bond – the force that holds atoms together in molecules and compounds.

Chemical energy – potential energy stored in chemical bonds of molecules that can be released by a chemical reaction.

Chernobyl – a city in Ukraine where a nuclear power plant suffered a meltdown due to poor decisions made by power plant workers; the resulting explosions killed some workers and leaked radioactive particles into the atmosphere.

Chlorinated hydrocarbon – an organic compound in which chlorine atoms have replaced most of the hydrogen atoms; also called an "organochloride," "organochlorine compound" or "chlorocarbon."

Chlorine – a halogen element that is isolated as a heavy greenish-yellow gas of pungent odor; used especially as a bleach, oxidizing agent and disinfectant in water purification.

Chlorophyll – any of a group of green pigments essential in photosynthesis.

Chloroplasts – chlorophyll-containing organelles in eukaryotic organisms; sites of photosynthesis.

Chronic effects – long-lasting results of exposure to a toxin; can be a permanent change caused by a single, acute exposure or, over time, by continuous, low-level exposure.

Chronic food shortages – long-term under-nutrition and malnutrition; usually caused by people's lack of money to buy food or lack of opportunity to grow it themselves.

Circular metabolism – a system in which wastes, especially water and materials, are reused and recycled; compare with *Linear metabolism*.

CITES Treaty – the *"Convention on International Trade in Endangered Species"* agreement among 167 governments aiming to ensure that cross-border trade in wild animals and plants does not threaten their survival; in 1989, participating countries agreed to ban all ivory trade.

Citizen science – projects in which trained volunteers work with scientific researchers to answer real-world questions.

City – a differentiated community with enough population and resource base to allow residents to specialize in arts, crafts, services, and professional occupations.

Civil law – a body of laws regulating relations between individuals or between individuals and corporations concerning property rights, personal dignity and freedom, and personal injury.

Class A pan – (water management) an open pan used as a standard for measuring water evaporation.

Classical economics – modern Western economic theories of the effects of resource scarcity, monetary policy and competition on the supply of and demand for goods and services in the marketplace; the basis for the capitalist market system.

Clay – a fine-grained, firm earthy material that is plastic when wet and hardens when heated, consisting primarily of hydrated silicates of aluminum; widely used in making bricks, tiles, and pottery, as well as liners in landfills because it is *Impervious*.

Clean Air Act – long-standing federal legislation that is the legal basis for the national clean air programs; last amended in 1990.

Cleaner production – the continual effort to prevent pollution, reduce the use of energy, water, and material resources, and minimize waste, all without reducing production capacity.

Clear-cut/Clearcutting – cutting every tree in a given area, regardless of species or size; an appropriate harvest method for some species; can be destructive if not carefully controlled.

Climate – the general variations of weather in a region over long periods (i.e., the "average weather"); see *Weather*.

Climate change – a change in weather over time and/or region; usually relating to changes in temperature, wind patterns, and rainfall; although it may be natural or anthropogenic, common discourse carries the assumption that climate change is anthropogenic.

Climax community – a relatively stable, long-lasting community arrived at through a series of succession; usually determined by climate and soil type.

Closed canopy – a forest where tree crowns spread over 20 percent of the ground; has the potential for commercial timber harvests.

Cloud forests – high mountain forests where temperatures are uniformly cool, and fog or mist keeps vegetation wet all the time.

Coal gasification – the heating and partial combustion of coal to release volatile gases, such as methane and carbon monoxide; after the pollutants are washed out, these gases become efficient, clean-burning fuel.

Coal liquefaction – a chemical process by which solid coal is converted to a liquid; referred to as a "synfuel" or "synthetic fuel."

Coal washing – coal technology that involves crushing coal and washing out soluble sulfur compounds with water or other solvents.

Coastal Zone Management Act – legislation of 1972 that gave federal money to 30 seacoasts and Great Lakes states for development and restoration projects.

Co-composting – microbial decomposition of organic materials in solid waste into useful soil additives and fertilizer; often, extra organic material in the form of sewer sludge, animal manure, leaves, and grass clippings are added to solid waste to speed the process and make the product more useful.

Coevolution – the process in which species exert selective pressure on each other and gradually evolve new features or behaviors as a result of those pressures.

Cogeneration – a power generation process that increases efficiency by harnessing the heat that would otherwise be wasted in the fuel combustion process and using it to generate electricity, warm buildings or for other purposes.

Cohousing – clusters of houses having shared dining halls and other spaces, encouraging stronger social ties while reducing the material and energy needs of the community.

Coir – the fiber of a coconut.

Cold front – a moving boundary of cooler air displacing warmer air.

Coliform bacteria – bacteria that live in the intestines (including the colon) of humans and other animals; used as a measure of the presence of feces in water or soil.

Combustion – a chemical change, especially oxidation, accompanied by the production of heat and light.

Command and control – require polluters to meet specific emission-reduction targets and often requires the installation and use of specific types of equipment to reduce emissions.

Commensalism – a symbiotic relationship in which one member is benefited, and the other is neither harmed nor benefited.

Commercial and industrial waste – (waste management) solid waste generated by the business sector as well as that created by state and federal governments, schools and tertiary institutions; does not include waste from the construction and demolition industry.

Commercial breeding – breeding animals or plants for commercial purposes, such as breeding dogs.

Commercial harvesting – the harvesting of animals or cash crops for commercial reasons.

Commingled materials – (waste management) materials mixed, such as plastic bottles, glass, and metal containers; commingled recyclable materials require sorting after collection before they can be recycled.

Common law – the body of court decisions that constitute a working definition of individual rights and responsibilities where there are no formal statutes that define these issues.

Communal resource management systems – resources directed by a group of people for long-term sustainability.

Community – a group of various populations in each area

Comparative risk assessment – a methodology which uses science, policy, economic analysis and stakeholder participation to identify and address areas of greatest environmental risk; a method for assessing environmental management priorities; the U.S. EPA offers free software which contains the history and methodology of comparative risk, as well as many case studies.

Compensation point – the point where the amount of energy produced by photosynthesis equals the amount of energy released by respiration.

Competitive exclusion – the theory that no two populations of different species will be able to occupy the same niche and compete for the same resources in the same habitat for very long.

Complexity (ecological) – the number of species at each trophic level and the number of trophic levels in a community.

Compost – the aerobically decomposed remnants of organic matter.

Composting – the biological decomposition of organic materials in the presence of oxygen that yields carbon dioxide, heat and stabilized organic residues that may be used as a soil additive.

Compound – a molecule made up of two or more kinds of atoms held together by chemical bonds.

Concentration – the amount of a component in each area or volume.

Condensation – 1) the change of state from a gas to a liquid (e.g., when water vapor in the air changes to liquid as it cools); 2) the aggregation of water molecules changing from vapor to liquid or solid when the saturation concentration is exceeded.

Condensation nuclei – tiny particles that float in the air and facilitate the condensation process.

Confined aquifer – aquifers that have the water table above their upper boundary and are typically found below unconfined aquifers.

Conifer – needle-bearing trees that produce seeds in cones.

Conservation development – consideration of landscape history, human culture, topography, and ecological values in subdivision design; using cluster housing, zoning, covenants, and other design features, at least half of a subdivision can be preserved as open space, farmland or natural areas.

Conservation of matter – in any chemical reaction, matter changes form; it is neither created nor destroyed.

Conspicuous consumption – a term coined by economist and social critic Thorstein Veblen to describe lavish spending on goods and services acquired mainly to display one's income or wealth rather than to satisfy the basic needs of the consumer.

Construction and demolition waste – (waste management) includes waste from residential, civil and commercial construction and demolition activities, such as fill material (e.g., soil), asphalt, bricks, and timber; excludes construction waste, which is included in the municipal waste stream; does not generally include waste from the commercial and industrial waste stream.

Consumer – 1) an organism that obtains energy and nutrients by feeding on other organisms or their remains; see *Heterotroph*; 2) an industry that maintains itself by transforming a high-quality energy source into a lower-quality one.

Consumer democracy – using your economic capacity to promote your values.

Consumption – the fraction of withdrawn water that is lost in transmission or that is evaporated, absorbed, chemically transformed, or otherwise made unavailable for other purposes as a result of human use.

Consumption (ecology) – the use of resources by a living system; the inflow and degradation of energy that is used for system activity.

Consumption (economics) – part of disposable income (income after taxes paid and payments received) that is not saved; essentially the goods and services used by households; includes purchased commodities at the household level (such as food, clothing and utilities), the goods and services paid for by government (such as defense, education, social services and healthcare) and the resources consumed by businesses to increase their assets (such as business equipment and housing).

Consumptive – of or pertaining to consumption; having the quality of consuming or dissipating; consumptive uses of water include pumping water for irrigation or municipal uses and *Evapotranspiration*.

Containment building – reinforced concrete building housing a nuclear reactor; designed to contain an explosion should one occur.

Contaminants – something that makes a place or substance no longer suitable for use.

Continental – relating to or characteristic of a continent (one of the large landmasses of the Earth).

Contour plowing (contour farming) – the farming practice of plowing across a slope following its contours; the rows formed have the effect of slowing water run-off during rainstorms so that the soil is not washed away and allowing the water to percolate into the soil.

Control rods – neutron-absorbing material inserted into spaces between fuel assemblies in nuclear reactors to regulate fission reaction.

Controlled burning – a technique sometimes used in forest management, farming, prairie restoration or greenhouse gas abatement.

Convection cell – the transfer of heat or other atmospheric properties by massive motion within the atmosphere, especially by such motion directed upward.

Convection currents – rising or sinking air currents that stir the atmosphere and transport heat from one area to another; also occur in water; see *Spring overturn*.

Convention on the International Trade in Endangered Species (CITES) – an International agreement among 167 governments aiming to ensure that cross-border trade in wild animals and plants does not threaten their survival; the species covered by CITES are listed in three Appendices, according to the degree of protection they need.

Conventional energy – energy from sources, such as fossil fuels, that are widely used.

Convert – to express a quantity in alternative units.

Cool deserts – deserts that are characterized by cold winters and sagebrush, such as the American Great Basin.

Coral reefs – prominent oceanic features composed of hard, lime skeletons produced by coral animals; usually formed along the edges of shallow, submerged ocean banks or shelves in warm, shallow tropical seas.

Core – the dense, intensely hot mass of molten metal, mostly iron and nickel that is thousands of kilometers in diameter at the Earth's center.

Core region – the primary industrial region of a country; usually located around the capital or largest port; has both the greatest population density and the greatest economic activity of the country.

Coriolis effect – the observed effect of the Coriolis force, especially the inertial force caused by the Earth's rotation in deflecting an object moving above the Earth, rightward in the northern hemisphere, leftward in the southern hemisphere.

Cornucopian fallacy – the belief that nature is limitless in its abundance and that perpetual growth is not only possible but essential.

Corporate Social Responsibility – integration of social and environmental policies into day-to-day corporate business.

Corridor – a strip of natural habitat that connects two adjacent nature preserves to allow migration of organisms from one place to another.

Corrosive – 1) causing damage to metal during chemical processes; 2) causing something to become weak; gradually destructive; steadily harmful.

Cost-benefit analysis – evaluation of large-scale public projects by comparing the costs and the benefits that will accrue from them.

Covenants – formal agreements or contracts, often between government and industry sectors; the national packaging covenant and sustainability covenants are examples of voluntary covenants with a regulatory underpinning; land covenants protect land for wildlife into the future.

Cover crops – plants, such as rye, alfalfa or clover, that can be planted immediately after a harvest to hold and protect the soil.

Cracking – the breaking of the long carbon chains found in the hydrocarbons in crude oil by heating them at high temperatures to form smaller molecules that are more useful.

Criminal law – a body of court decisions based on federal and state statutes concerning wrongs against persons or society.

Criteria pollutants – the 1970 amendments to the Clean Air Act required EPA to set National Ambient Air Quality Standards for certain pollutants known to be hazardous to human health; the EPA has identified six criteria pollutants: sulfur dioxide, carbon monoxide, lead, nitrogen oxides, ozone, and particulate matter.

Critical factor – the single environmental factor closest to the tolerance limit for a given species at a given time; see *Limiting factors*.

Critical thinking – an ability to evaluate information and opinions in a systematic, purposeful, efficient manner.

Crop coefficient (Kc) – (water management) a variable used to calculate the *Evapotranspiration* of a plant crop based on that of a reference crop.

Crop evapotranspiration – (water management) the crop water use (i.e., the daily water withdrawal).

Crop rotation (crop sequencing) – the practice of growing dissimilar types of crops in the same space in sequential seasons for various benefits, such as to avoid the buildup of pathogens and pests that often occurs when one species is continuously cropped.

Croplands – land that is suitable to grow crops.

Crude death rate – the number of deaths per thousand persons in a given year; also called the "crude mortality rate."

Crude oil – naturally occurring mixture of hydrocarbons under normal temperature and pressure.

Crust – the cool, lightweight, outermost layer of the Earth's surface that floats on the soft, pliable underlying layers.

Curbside collection – a collection of household recyclable materials (separated or co-mingled) that are left at the curbside for collection by local government services.

Cullet – crushed glass that is suitable for recycling by glass manufacturers.

Cultural eutrophication – an increase in biological productivity and ecosystem succession caused by human activities.

Cultural services – the non-material benefits of ecosystems, including refreshment, spiritual enrichment, knowledge and artistic satisfaction.

Culture jamming – altering existing mass media to criticize itself (e.g., defacing billboards with an alternative message); public activism opposing commercialism as little more than propaganda for established interests and the attempt to find alternative expression.

Culvert – drain that passes under a road or pathway, may be a pipe or other conduit.

Cut and fill – removing earth from one place to another, usually mechanically.

Cyanobacteria (Cyanophyta or blue-green algae) – a phylum of bacteria that obtain their energy through photosynthesis.

Cyclone – intense, low-pressure weather systems; mid-latitude cyclones are atmospheric circulations that rotate clockwise in the Southern Hemisphere and anti-clockwise in the Northern Hemisphere; generally associated with stronger winds, unsettled conditions, cloudiness, and rainfall; tropical cyclones, called *Hurricanes* in the Northern Hemisphere, cause storm surges in coastal areas.

D

Daughter – a material formed from the parent material after a given process, such as nuclear decay or movement through the rock cycle.

DDT (dichlorodiphenyltrichloroethane) – a colorless, odorless water-insoluble crystalline insecticide $C_{14}H_9C_{15}$ that tends to accumulate in ecosystems and has toxic effects on many vertebrates; became the most widely-used pesticide from WWII to the 1950s; implicated in illnesses and environmental problems; now banned in the U.S.

Debt-for-nature swap – a financial transaction in which a portion of a developing nation's foreign debt is forgiven in exchange for local investments in conservation measures.

Deciduous – trees and shrubs that shed their leaves at the end of the growing season.

Deciduous forest – a forest made up of trees that drop their leaves seasonally.

Decline spiral – a catastrophic deterioration of a species, community or whole ecosystem; accelerates as functions are disrupted or lost in a downward cascade.

Decomposers – consumers, mostly microbial, that change dead organic matter into minerals and heat.

Deductive reasoning – deriving testable predictions about specific cases from general principles.

Deep ecology – a philosophy that calls for a profound shift in our attitudes and behavior based on voluntary simplicity, rejection of anthropocentric attitudes, intimate contact with nature, decentralization of power, support for cultural and biological diversity, a belief in the sacredness of nature and direct personal action to protect nature, improve the environment and bring about fundamental societal change.

Deforestation – the conversion of forested areas to non-forest land for agriculture, urban use, development or wasteland.

Degradation (of water resource) – deterioration in water quality due to contamination or pollution; makes the water unsuitable for other desirable purposes.

Delaney Clause – a controversial amendment to the Federal Food, Drug and Cosmetic Act added in 1958 prohibiting the addition of any known cancer-causing agent to processed foods, drugs or cosmetics.

Delivered energy – energy delivered to and used by a household; usually gas and electricity.

Delta – fan-shaped sediment deposit found at the mouth of a river.

Demand – the amount of a product that consumers are willing and able to buy at various possible prices, assuming they are free to express their preferences.

Demanufacturing – disassembly of products so components can be reused or recycled.

Dematerialization – decreasing the consumption of materials and resources while maintaining the quality of life.

Demographic transition – 1) a change in the make-up of a human population or group from one set of characteristics to another; 2) a pattern of falling death rates and, often, birthrates in response to improved living conditions; could be reversed in deteriorating conditions.

Demographics – the characteristics of a human population or a part of it, especially its size, growth, density, distribution and statistics regarding birth rate, marriage, the incidence of disease and death rate.

Demography – 1) vital statistics about people: births, marriages, deaths, etc.; 2) the statistical study of human populations relating to growth rate, age structure, geographic distribution, etc. and their effects on social, economic and environmental conditions.

Denitrifying bacteria – free-living soil bacteria that convert nitrates to gaseous nitrogen and nitrous oxide.

Density – the quantity of something per unit measure, especially per unit length, area or volume; the mass per unit volume of a substance under specified conditions of pressure and temperature.

Dependency ratio – the number of non-working members, compared to working members for a given population.

Depository or repository – a place where something is kept for safekeeping or storage, such as a warehouse or store for furniture or valuables; Yucca Mountain in New Mexico is being studied as a potential depository for spent nuclear fuel.

Desalination – producing potable or recyclable water by removing salts from salty or brackish water; three methods do this: (1 distillation/freezing (2 reverse osmosis using membranes and electrodialysis and (3 ion-exchange; at present, all these methods are energy-intensive.

Desert – an area that receives average annual precipitation of less than 250 mm (9.8 inches) or an area in which more water is lost than falls as precipitation.

Desertification – denuding and degrading a once-fertile land, initiating a desert-producing cycle that feeds on itself and causes long-term changes in the soil, climate, and *Biota* of an area.

Detritivore – an organism that consumes organic litter, debris, and dung and in so doing contributes to decomposition and the recycling of nutrients.

Detritus – non-living particulate organic material (as opposed to dissolved organic material).

Detritus feeders – organisms that obtain their nutrients and energy by breaking down dead materials and organic compounds in an ecosystem.

Developing countries – development of a country is measured using a mix of economic factors (income per capita, GDP, degree of modern infrastructure (both physical and institutional), degree of industrialization (the proportion of the economy devoted to agriculture compared to that devoted to natural resource extraction) and social factors (life expectancy, the rate of literacy, poverty); the UN-produced Human Development Index (HDI) is a compound indicator of the above statistics; there is a strong correlation between low-income and high population growth, both within and between countries; in developing countries, there is low per capita income, widespread poverty, and low capital formation; in developed countries, there is continued economic growth and a relatively high standard of living; the term is value-laden and prescriptive, as it implies a natural transition from "undeveloped" to "developed," although such transitions can be imposed or coaxed along; although poverty and physical deprivation are clearly undesirable, it does not follow that it is therefore desirable for "undeveloped" economies to be coaxed toward affluent Western-style "developed," free market economies; the terms "industrialized" and "non-industrialized" are no different in this assumption.

Dewpoint – the temperature at which condensation occurs for a given concentration of water vapor in the air.

dfE – an acronym for "design for the environment;" dfE considers "cradle to grave" costs and benefits associated with the material acquisition, manufacture, use, and disposal.

dfM – acronym for "design for manufacturing;" designing products in such a way that they are easy to manufacture.

dfS – acronym for "design for sustainability;" an integrated design approach aiming to achieve both environmental quality and economic efficiency through the redesign of industrial systems.

dfX – designing for assembly/disassembly, re-use, or recycling.

Dieback – 1) in arboriculture, a condition in trees or woody plants in which peripheral parts are killed, either by parasites or due to conditions such as acid rain; 2) a sudden population decline; see *Population crash*.

Diesel – 1) fuel that is made of hydrocarbons that are 16 carbons long; 2) a high-compression internal combustion engine.

Dietary energy supply – food available for human consumption, usually expressed in kilocalories per person per day.

Diminishing returns – a condition in which unrestrained population growth causes the standard of living to decrease to a subsistence level where poverty, misery, vice, and starvation makes life permanently drab and miserable; this dreary prophecy has led economics to be called "the dismal science."

Dioxin – any one of some chemical compounds that are persistent organic pollutants and are carcinogenic.

Direct action – civil disobedience, guerrilla street theater, picketing, protest marches, road blockades, demonstrations and other techniques borrowed from the civil rights movement and applied to environmental protection.

Direct energy – the energy being currently used, used mostly at home (delivered energy) and for fuels used mainly for transport.

Disability-adjusted life years (DALY) – a measure of premature deaths and losses due to the onset of illnesses and disabilities in a population.

Discharge – the amount of water that passes a fixed point in a given amount of time, usually expressed as liters or cubic feet of water per second.

Discharge rate – the amount of water that passes a fixed point in a given amount of time, usually expressed as liters or cubic feet of water per second.

Disease – a deleterious change in the body's condition in response to destabilizing factors, such as nutrition, chemicals or biological agents.

Disinfection – to make free from infection by destroying harmful microorganisms.

Dissemination – to become widely scattered (seeds).

Dissolved Oxygen (DO) content – the amount of oxygen dissolved in a given volume of water at a given temperature and atmospheric pressure, usually expressed in parts per million.

Distillation – 1) the extraction of volatile components of a mixture by the condensation and collection of the vapors that are produced as the mixture is heated; 2) a process of desalinization in which water is evaporated and then re-condensed.

Distributed water – (water management) purchased water supplied to a user; this is usually through a reticulated mains system (but also through pipes and open channels, irrigation systems supplied to farms).

Diversion rate – (waste disposal) the proportion of a potentially recyclable material that has been diverted out of the waste disposal stream and, therefore, not directed to landfills.

Diversity (species diversity, biological diversity) – the number of species present in a community (species richness), as well as the relative abundance of each species.

Divertible resource – (water management) the proportion of water runoff and recharge that can be accessed for human use.

DNA (Deoxyribonucleic acid) – the long, double-helix molecule in the nucleus of cells that contains the genetic code and directs the development and functioning of all cells.

Dominant – 1) an organism that behaves in such a way as to be in a position over others of the same species; 2) the allele of a gene that requires only one copy to be present in an individual for that trait to be present.

Dominant plants – those plant species in a community that provide the food base for most of the community; usually take up the most space and have the largest biomass.

Dose threshold level – the maximum level of a substance before toxic levels are reached.

Downbursts – sudden, very strong downdrafts of cold air associated with an advancing storm front.

Downcycling – (waste management) recycling in which the quality of an item is diminished with each recycling.

Downstream – those processes occurring after an activity (e.g., the transport of a manufactured product from a factory to the wholesale or retail outlet); see *Upstream*.

Drainage – (water management) that part of irrigation or rainfall that runs off an area or is lost to deep percolation.

Drawdown – (water management) a drop in the water level, generally applied to wells or bores.

Dredging – (water management) the repositioning of soil from an aquatic environment using specialized equipment to initiate infrastructural or ecological improvements.

Drift net – a type of fishing net used in oceans, coastal seas, and freshwater lakes.

Drinking water – (potable water) water fit for human consumption by the *World Health Organization* (WHO) guidelines.

Drip irrigation – (water management) using pipe or tubing perforated with very small holes to deliver water one drop at a time directly to the soil around each plant; conserves water, prevents soil waterlogging, and reduces the salt content.

Driver – (ecology) any natural or human-induced factor that directly or indirectly causes a change in an ecosystem; a "direct driver" is one that unequivocally influences ecosystem processes, and that can be measured.

Drop-off center – (waste management) a location where discarded materials can be left for recycling.

Drought – 1) an acute water shortage relative to availability, supply, and demand in a region; 2) an extended period of months or years when a region notes a deficiency in its water supply; generally, this occurs when a region has received below-average precipitation consistently.

Drought cycle – temporary yet repetitive phases of dry conditions in an otherwise hospitable environment.

Dry alkali injection – spraying dry sodium bicarbonate into flue-gas to absorb and neutralize acidic sulfur compounds; see *Flue-gas scrubbing*.

Dryland salinity – (water management) accumulation of salts in soils, soil water, and groundwater; may be natural or induced by land clearing

Dung – animal excrement (biomass); used as fuel for heating or cooking in many countries.

Dynamic state of equilibrium – a steady-state found in an ecosystem or a system where change is not observable because, even though there are changes in progress, they are being made at an equal rate with no net gain.

E

Earth charter – a set of principles for sustainable development, environmental protection, and social justice developed by a council appointed by the United Nations.

Earthquakes – a sudden, violent movement of the Earth's crust.

Eco- – representing "ecology;" a prefix now added to many words indicating a consideration for the environment (e.g., eco-housing, eco-label, eco-material).

Eco-asset – a biological asset that provides financial value to private landowners when it is maintained in or restored to its natural state.

Ecocentric (ecologically centered) – a philosophy that claims moral values and rights for both organisms and ecological systems and processes.

Ecofeminism – a pluralistic, nonhierarchical, relationship-oriented philosophy that suggests how humans could reconceive themselves and their relationships to nature in non-dominating ways (devised as an alternative to patriarchal systems of domination).

Ecojustice – justice in the social order and integrity in the natural order.

Ecolabel – seal or logo indicating a product has met certain environmental or social standards.

Ecological deficit – of a country or region; measures the amount by which its *Ecological Footprint* exceeds the ecological capacity of that region.

Ecological development – a gradual process of environmental modification by organisms.

Ecological economics – application of ecological insights to economic analysis in a holistic, contextual, value-sensitive, eco-centric manner.

Ecological equivalents – different species that occupy similar ecological niches in similar ecosystems in different parts of the world.

Ecological Footprint (Eco-footprint, Footprint) – a measure of the area of biologically productive land and water needed to produce the resources and absorb the wastes of a population using the prevailing technology and resource management schemes; a measure of the consumption of renewable natural resources by a human population, be it that of a country, a region or the whole world; given as the total area of productive land or sea required to produce all the crops, meat, seafood, wood and fiber that population consumes to sustain its energy consumption and to provide space for its infrastructure.

Ecological niche – 1) the habitat of a species or population within its ecosystem; 2) the functional role and position of a species (population) within a community or ecosystem, including what resources it uses, how and when it uses the resources and how it interacts with other populations.

Ecological succession – 1) the process by which organisms occupy a site and gradually change environmental conditions so that other species can replace the original inhabitants; 2) the predictable and orderly changes in the composition or structure of an ecological community over time.

Ecological sustainability – the capacity of ecosystems to maintain their essential processes and function and to retain their biological diversity without impoverishment.

Ecologically sustainable development – using, conserving and enhancing the human community's resources so that ecological processes, on which all life depends, can be maintained and enriched into the future.

Ecology – 1) the scientific study of living organisms and their relationships to one another; concerned with the life histories, distribution, and behavior of individual species as well as the structure and function of natural systems at the level of populations, communities, and ecosystems; 2) the scientific study of the processes regulating the distribution and abundance of organisms.

Economic development – a rise in real income per person; usually associated with new technology that increases productivity or resources.

Economic globalization – the emerging international economy, characterized by free trade in goods and services, unrestricted capital flows, and more limited national powers to control domestic economies.

Economic growth – an increase in the total wealth of a nation; if the population grows faster than the economy, there may be macroeconomic growth, but the share per person may decline.

Economic thresholds – in pest management, the point at which the cost of pest damage exceeds the costs of pest control.

Ecoregion – the next-smallest ecologically and geographically defined area beneath "realm" or "ecozone," which are divisions of the Earth's surface defined by the distribution of organisms; see *Bioregion*.

Ecosystem – a dynamic complex of plant, animal and microorganism communities and their non-living environment, all interacting as a functional unit.

Ecosystem boundary – the spatial delimitation of an ecosystem, usually based on discontinuities of organisms and the physical environment.

Ecosystem management –integration of ecological, economic and social goals in a unified systems approach to resource management.

Ecosystem restoration – to reinstate an entire community of organisms to as near its natural condition as possible.

Ecosystem services – the role played by organisms, without charge, in creating a healthy environment for human beings, from the production of oxygen to soil formation, maintenance of water quality and much more; these services are now generally divided into four groups: supporting, provisioning, regulating and cultural.

Ecotage – direct action (guerrilla warfare) or sabotage in defense of nature; see *Monkeywrenching*.

Ecotone – a boundary between two types of ecological communities.

Ecotourism – a combination of adventure travel, cultural exploration and nature appreciation in wild settings.

E-cycling – electronic recycling waste.

Edge effects – a change in species composition, physical conditions, or other ecological factors at the boundary between two ecosystems.

Effective rainfall – the volume of rainfall passing into the soil; that part of rainfall available for plant use after runoff, leaching, evaporation and foliage interception.

Effluent – a discharge or emission of liquid, gas or other waste product.

Effluent sewerage – a low-cost alternative sewage treatment for cities in developing countries that combines some features of septic systems and centralized municipal treatment systems.

El Niño – a warm water current which periodically flows southwards along the coast of Ecuador and Peru in South America, replacing the usually cold northward-flowing current; occurs once every five to seven years, usually during the Christmas season (the name refers to the Christ child); the opposite phase is called *La Niña*.

El Niño Southern Oscillation (ENSO) – the formation of an El Niño is linked with the cycling of a Pacific Ocean circulation pattern known as the southern oscillation; in a normal year, a surface low pressure develops in the region of northern Australia and Indonesia and a high-pressure system over the coast of Peru; the trade winds over the Pacific Ocean move strongly from east to west; the easterly flow of the trade winds carries warm surface waters westward, bringing convective storms to Indonesia and coastal Australia; along the coast of Peru, cold bottom water wells up to the surface to replace the warm water that is pulled to the west.

Electron – a negatively charged subatomic particle that orbits around the nucleus (protons and neutrons) of an atom.

Electrostatic – of or relating to electric charges at rest or produced or caused by such charges.

Electrostatic precipitators – the most common particulate controls in power plants; fly ash particles pick up an electrostatic surface charge as they pass between large electrodes in the effluent stream, causing particles to migrate to the oppositely charged plate.

Element – a molecule composed of one kind of atom; cannot be broken into simpler units by chemical reactions.

Embodied energy – 1) the energy expended over the entire lifecycle of a good or service; 2) the energy involved in the extraction of basic materials, processing/manufacture, transport, and disposal of a product; 3) the energy required to provide a good or service.

Embodied water – the hidden flow of water that accompanies the trade of goods; see *Virtual water*.

Emergent diseases – a new disease or one that has been absent for at least 20 years.

Emergent property – a property that has not hitherto been evident in the individual components of an object or system.

Emergy – short for "energy memory;" all the available energy that was used in the work of making a product directly and indirectly, expressed in units of one type of available energy (the work previously done to provide a product or service); the energy of one type required to make energy of another.

Emigration – the movement of members from a population.

Emissions – substances, such as gases or particles, discharged into the atmosphere as a result of natural processes of human activities, including those from chimneys, elevated point sources, and tailpipes of motor vehicles.

Emission standards – regulations for restricting the amounts of air pollutants that can be released from specific point sources.

Emissions intensity – emissions expressed as quantity per monetary unit.

Emissions trading – a system by which states and institutions receive permits to produce a specified amount of carbon dioxide and other greenhouse gases, which they may trade with others; also called "*carbon trading*."

Endangered species – a species that is at risk of becoming extinct because it is either few in number or threatened by changing environmental or predation parameters.

Endemism – a state in which species are restricted to a single region.

Energetics – the study of how energy flows within an ecosystem: the routes it takes, rates of flow, where it is stored and how it is used.

Energy – 1) the capacity to do work (i.e., to change the physical state or motion of an object); 2) a property of all systems which can be turned into heat and measured in heat units; types include: *Embodied energy, Geothermal energy, Hydro energy, Kinetic* *energy, Nuclear energy, Potential energy, Primary energy, Solar energy, Secondary energy, Stationary energy, Tidal energy, Useful energy, Wind energy, et al.*

Energy accounting – measuring value by the energy input required for a good or service; a form of accounting that builds in a measure of our impact on nature, rather than being restricted to human-based items.

Energy audit – a systematic gathering and analysis of energy use information that can be used to determine energy efficiency improvements; the Australian and New Zealand Standard AS/NZS 3598:2000 Energy Audits defines three levels of audit for consistency: the basic, detailed and precision energy audits.

Energy crisis – a significant rise in price due to the scarcity of energy supplies within an economy.

Energy cycle – how energy is cycled through the biosphere (e.g., when the sun's energy is taken up by plants, from plants to animals and from animals to other animals through ingestion).

Energy efficiency – using less energy to provide the same level of energy service; a measure of energy produced compared to the energy consumed.

Energy footprint – the area required to provide or absorb the waste from coal, oil, gas, fuelwood, nuclear energy, and hydropower; The Fossil Fuel Footprint is the area required to sequester the emitted CO_2, considering CO_2 absorption by the sea, etc.

Energy-for-land ratio – the amount of energy that can be produced per hectare of ecologically productive land. The units used are gigajoules per hectare/year, or GJ/ha/yr; for fossil fuel, calculated as CO_2 assimilation, the ratio is 100 GJ/ha/yr.

Energy management – A program of well-planned actions aimed at reducing energy use, recurrent energy costs, and detrimental greenhouse gas emissions.

Energy pyramid – a representation of the loss of useful energy at each step in a food chain.

Energy recovery – 1) incineration of solid waste to produce useful energy; 2) the productive extraction of energy, usually electricity or heat, from waste or materials that would otherwise have gone to landfill.

Enhanced greenhouse effect – the increase in the natural greenhouse effect resulting from increases in atmospheric concentrations of greenhouse gases due to emissions from human activities.

Enriched uranium – uranium (U) ore occurs naturally in a state that cannot be used in most reactors or to make nuclear weapons; makes it easier to use in reactors; the enrichment process increases the amount of the fissionable ^{235}U isotope; uranium enriched to contain less than 20% ^{235}U is called low-enriched uranium; uranium enriched to contain 20% or greater ^{235}U is highly-enriched uranium that can be directly used to make nuclear weapons.

Entropy – for a closed thermodynamic system, a quantitative measure of the amount of thermal energy not available to do work; the symbol is S.

Environment – the circumstances or conditions that surround an organism or group of organisms as well as the complex of social or cultural conditions that affect an individual or community.

Environmental ethics – a search for moral values and ethical principles in human relations with the natural world.

Environmental flows – river or creek water flows that are allocated for the maintenance of the waterway ecosystems.

Environmental hormones – chemical pollutants that, over time, come to substitute for, or interfere with, naturally occurring hormones in our bodies; these chemicals may trigger a reproductive failure, developmental abnormalities, or tumor promotion.

Environmental impact statement (EIS) – an analysis, required by provisions in the *National Environmental Policy Act* of 1970, of the effects of any major program a federal agency plans to undertake.

Environmental indicator – any physical, chemical, biological, or socio-economic measure that can be used to assess natural resources and environmental quality; more specifically, organisms with these characteristics are called "bioindicators."

Environmental justice – a recognition that access to a clean, healthy environment is a fundamental right of all human beings.

Environmental law – the special body of official rules, decisions, and actions concerning environmental quality, natural resources, and ecological sustainability.

Environmental literacy – fluency in the principles of ecology that gives us a working knowledge of the basic grammar and underlying syntax of environmental wisdom.

Environmental movement (environmentalism) – both the conservation and green movements; a diverse scientific, social and political movement; in general terms, environmentalists advocate the sustainable management of resources and stewardship of the natural environment through changes in public policy and individual behavior; in its recognition of humanity as a participant in ecosystems, the movement is centered around ecology, health, and human rights.

Environmental policy – the official rules or regulations concerning the environment adopted, implemented and enforced by some governmental agency.

Environmental racism – decisions that restrict certain groups of people to polluted or degraded environments by race.

Environmental resistance – all the limiting factors that tend to reduce population growth rates and set the maximum allowable population size or carrying capacity of an ecosystem; compare with *Biotic potential*.

Environmental resources – anything an organism needs that can be provided by the environment.

Environmental science – the study of interactions among physical, chemical and biological components of the environment, as well as our role in it.

Environmentalism – active participation in attempts to solve environmental pollution and resource problems.

Enzymes – molecules, usually proteins or nucleic acids, that act as catalysts in biochemical reactions.

Epidemiology – the study of factors affecting the health and illness of populations; serves as the foundation and logic of interventions made in the interest of public health and preventive medicine.

Epiphyte – a plant that grows on a substrate other than the soil, such as the surface of another organism.

Equilibrium community – a community subject to periodic disruptions, usually by fire, that prevent it from reaching a climax stage; also called a "*disclimax community*."

Erosion – displacement of solids, such as sediment, soil, rock, and other particles, usually by the agents of currents such as wind, water or ice by downward or down-slope movement in response to gravity or by living organisms; removal of vegetation and trees can increase erosion of topsoil.

***Escherichia coli* (*E. Coli*)** – a bacterium used as an indicator of fecal contamination and potential disease organisms in the water.

Estimated reserves – reserves of resources whose quantities have only been estimated and are not known for certain.

Estuary – 1) the wide lower course of a river where the tide flows in, causing fresh and saltwater to mix; 2) a semi-enclosed coastal body of water with one or more rivers or streams flowing into it and with a free connection to the open sea.

Ethical consumerism – buying things that are made ethically (i.e., without harm to or exploitation of humans, animals or the natural environment); generally, entails favoring products and businesses that take account of the greater good in their operations.

Ethical living – adopting lifestyles, consumption, and shopping habits that minimize our negative impact and maximize our positive impact on people, the environment, and the economy; see *Consumer democracy*.

Eukaryotic cell – a cell containing a membrane-bounded nucleus and membrane-bounded organelles.

Eutectic fluid – eutectic salts (salts that melt at low temperatures) are phase-changing chemicals that are used in active solar heating to store solar energy; heating melts these materials and cooling returns them to the original phase.

Eutrophic – rivers and lakes rich in organisms and organic material, often due to runoff from land (eu = well; trophic = nutritious).

Eutrophication – the enrichment of water bodies with nutrients, primarily nitrogen and phosphorus, which stimulate the growth of aquatic organisms.

Euxinic – with extremely low oxygen; see *Anoxic*.

Evaporation – the process by which liquid is changed into vapor (the gas phase) due to an increase in temperature or pressure.

Evapotranspiration (ET) – the sum of water evaporation and plant transpiration; actual evapotranspiration can never be greater than precipitation and will usually be less because some water will run off in rivers and flow to the oceans.

Evergreen – coniferous trees and broad-leaved plants that retain their leaves year-round.

Evolution – a theory that explains how random changes in genetic material and competition for scarce resources cause species to change gradually.

E-waste – electronic waste, especially mobile phones, televisions, and personal computers.

Exhaustible resources – generally considered the Earth's geologic endowment: minerals, non-mineral resources, fossil fuels, and other materials present in finite amounts in the environment.

Existence value – an economic value in which the benefit of just knowing that a particular species, organism or resource exists is appraised.

Exotic organisms – alien species introduced by the human agency into biological communities where they would not naturally occur.

Exponential curve – describes growth at a constant rate of increase per unit of time; it can be expressed as a constant fraction or exponent.

Exponential growth – growth at a constant rate of increase per unit of time; can be expressed as a constant fraction or exponent; see *Geometric growth*.

Extended producer responsibility (EPR) (product take-back) – a requirement (often in law) that producers take back and accept accountability for the responsible disposal of their products; this encourages the design of products that can be easily repaired, recycled, reused or upgraded.

External costs – expenses, monetary or otherwise, borne by someone other than the individuals or groups who use a resource.

External water footprint – the embodied water of imported goods; compare with *Internal water footprint*.

Externality – in environmental economics, by-products of activities that affect the well-being of people or damage the environment, where those impacts are not reflected in market prices; the costs or benefits associated with externalities do not enter standard cost accounting schemes; in many environmental situations environmental deterioration may be caused by a few while the cost is borne by the community (e.g., overfishing, the production of greenhouse emissions that are not compensated for in any way by taxes, etc.); the environment is often cited as a negatively affected externality of the economy.

Extinction – the irrevocable elimination of species; it can be a normal process of the natural world as species out-compete or kill off others or as environmental conditions change; reduction in biodiversity.

Extinction event (mass extinction, extinction-level event, ELE) – a sharp decrease in the number of species in a relatively short period.

Extirpate – to destroy a species totally; extinction caused by direct human action, such as hunting, trapping, etc.

Eye, Eye wall – the eye of a hurricane is its center, where no storm activity takes place; the eye wall is the area between the eye and the storm.

F

Fair trade – a guarantee that a fair price is paid to producers of goods or services; includes a range of other social and environmental standards, including safety standards and the right to form unions.

Family planning – controlling reproduction; planning the timing, and the number of births to have only as many babies as can be supported.

Famines – acute food shortages characterized by large-scale loss of life, social disruption, and economic chaos.

Fauna – all the animals present in a region.

Feces – food waste discharged from the bowels after it has been digested.

Fecundity – the physical ability to reproduce.

Feedback – flow from the products of action back to interact with the action.

Feedlot (feed yard) – a type of *Confined Animal Feeding Operation* (CAFO) that is used for fattening up livestock, notably beef cattle, before slaughter; also known as "factory farming."

Fen – an area of waterlogged soil that tends to be peaty; fed mainly by upwelling water; low productivity.

Feral – 1) in a natural, wild state; 2) a once domestic animal that has taken up a wild existence.

Fermentation – any of a group of chemical reactions induced by living or nonliving ferments that split complex organic compounds into relatively simple substances, especially the anaerobic conversion of sugar to carbon dioxide and alcohol by yeast.

Fertigate – to apply fertilizer through an irrigation system.

Fertility – a measurement of an actual number of offspring produced through sexual reproduction; usually described regarding the number of offspring of females, since paternity can be difficult to determine.

Fertility rate – the number of live births per 1,000 women aged 15 to 44 years; see *Birth rate* and *Mortality rate*.

Fertilization – the process of union of two gametes whereby the somatic chromosome number is restored, and the development of a new individual are initiated; the addition of materials to the soil to increase the available nutrient content.

Fertilizers – compounds given to plants to promote growth; they are usually applied either through the soil for uptake by plant roots or by foliar feeding for uptake through leaves.

Fetal alcohol syndrome – a tragic set of permanent physical, mental and behavioral congenital disabilities that can result when the mother drinks alcohol during pregnancy.

Fibrosis – the general name for the accumulation of scar tissue in the lung.

Fidelity – a principle that forbids misleading or deceiving any creature capable of being misled or deceived.

Filters – a porous mesh of cotton cloth, spun glass fibers or asbestos-cellulose that allows air or liquid to pass through but holds back solid particles.

Fire-climax community – an equilibrium community maintained by periodic fires (e.g., grasslands, chaparral shrubland, and some pine forests).

First law of thermodynamics – energy is conserved; energy is neither created nor destroyed; it is only changed to a different state (e.g., from chemical energy such as gasoline, into mechanical energy to power an engine and finally into heat from friction).

Flood control devices – measures to protect areas that are easily flooded by either reducing flood flows or confining the flow; devices include building dams or levees or modifying the channel of the river or stream.

Flood Disaster Protection Act of 1973 – this law signaled a shift in federal policy from reducing the floods through structural controls to reducing the damages by limiting the development in flood-prone areas by making federally-subsidized flood insurance available to property owners in flood-prone areas only in those communities which adopted *Floodplain* zoning.

Floodplains – low lands along riverbanks, lakes, and coastlines subjected to periodic inundation.

Flora – all of the plants present in a given region.

Fluctuations – rising and falling, such as population numbers; a variant.

Flue-gas scrubbing – treating combustion exhaust gases with chemical agents to remove pollutants; spraying crushed limestone and water into the exhaust gas stream to remove sulfur is a common scrubbing technique.

Fluidized bed combustion – high-pressure air is forced through a mixture of crushed coal and limestone particles, lifting the burning fuel and causing it to move like a boiling fluid.

Flyway – the flight paths used in bird migration; generally, span over continents and often oceans.

Food aid – financial assistance intended to boost less-developed countries' standards of living.

Food chain – a linked feeding series; in an ecosystem, the sequence of organisms through which energy and materials are transferred, in the form of food, from one trophic level to another; also called "food webs," "food networks" or "trophic networks."

Food miles – the emissions produced, and resources needed to transport food and drink around the globe.

Food security – food produced in sufficient quantity to meet the full requirements of all people (i.e., when the total global food supply equals the total global demand); for households it is the ability to purchase or produce the food they need for a healthy and active life (disposable income is a crucial issue); for national food security, the focus is on sufficient food for all people in a nation, and it entails a combination of national production, imports, and exports; food security always has components of production, access, and utilization.

Food surpluses – excess food supplies.

Food web – a complex, interlocking series of individual food chains in an ecosystem.

Footprint – as in *Ecological Footprint*, a measure of environmental impact, usually expressed as an area of productive land (the footprint) needed to counteract the impact.

Forage – the plant material (mainly plant leaves) eaten by grazing animals.

Forest – land with a canopy cover greater than 30%.

Forest management – scientific planning and administration of forest resources for sustainable harvest, multiple uses, regeneration, and maintenance of a healthy biological community.

Formula for photosynthesis – CO_2 (from the air) + H_2O + sun's energy (light) * $C_6H_{12}O_6$ (glucose) + O_2

Fossil fuel – any hydrocarbon deposit that can be burned for heat or power (and producing carbon dioxide when burnt); fuels formed from once-living organisms that have become fossilized over geological time (e.g., petroleum, natural gas, coal).

Fossil water – groundwater that has remained in an aquifer for thousands or millions of years; when geologic changes seal the aquifer, preventing further replenishment, the water becomes trapped inside and is then referred to as fossil water; a limited resource and can only be used once.

Fourth World – a political/economic category describing very developing nations that have neither market economies nor central planning and are either not developing or are developing very slowly; also used to describe poor indigenous communities within wealthier nations.

Freegan – a person using alternative strategies for living based on limited participation in the conventional economy and minimal consumption of resources; compounded from "free" and "vegan;" embrace community, generosity, social concern, freedom, cooperation and sharing; the most notorious freegan strategy is "urban foraging" or "dumpster diving," which is rummaging through the garbage of retailers, residences, offices and other facilities for useful goods; see *Affluenza* and *Froogle*.

Freezing condensation – a process that occurs in clouds when ice crystals trap water vapor; as the ice crystals become larger and heavier, they begin to fall as rain or snow.

Freon – DuPont's trade name for its odorless, colorless, nonflammable and noncorrosive chlorofluorocarbon and hydrochlorofluorocarbon refrigerants, used in air conditioning and refrigeration systems.

Freshwater – 1) water containing no significant amounts of salt; see *Potable water*; 2) water other than seawater; covers only about 2% of the Earth's surface, including streams, rivers, lakes, ponds, and water associated with several kinds of wetlands.

Freshwater ecosystems – ecosystems in which the fresh (non-salty) water of streams, rivers, ponds, or lakes plays a defining role.

Friction – the rubbing of two objects against each other when one or both is moving; a significant percentage of the energy produced by an automobile engine is dissipated in friction, reducing the overall efficiency of the system.

Front – in weather, the boundary between warm (high-pressure) and cold (low-pressure) air masses.

Froogle – a play on the word frugal; people who lead low-consumption lifestyles or are part of a new movement towards self-sufficiency and waste-reduction achieved by bartering goods and services (especially through the internet), making their own products, soap, clothes, breeding chickens and goats, growing their own food, baking their own bread, harvesting their own water and energy and helping to develop a sense of community; sometimes referring to people who have made a resolution to only buy essentials for a particular period of time; see *Freegan* and *Affluenza*.

Fuel assembly – a bundle of hollow metal rods containing uranium oxide pellets; used to fuel a nuclear reactor.

Fuel cell – an electrochemical device with no moving parts that converts the chemical energy of a fuel, such as hydrogen, and an oxidant, such as oxygen, directly into electricity; clean, quiet and highly efficient sources of electricity.

Fuel-switching – a change from one fuel to another.

Fuelwood – branches, twigs, logs, wood chips, and other wood products harvested for use as fuel.

Fugitive emissions – 1) in the context of the National Greenhouse Gas Inventory, these are greenhouse gases emitted from fuel production itself, including processing, transmission, storage, and distribution processes, and including emissions from oil and natural gas exploration, venting and flaring, as well as the mining of black coal; 2) substances that enter the air without going through a smokestack, such as dust from soil erosion, strip mining, rock crushing, construction and building demolition.

Fujita Scale – a scale measuring the intensity of a tornado based on its wind speed, diameter and the amount of damage caused.

Full-cost pricing – the pricing of commercial goods, such as electric power, that includes not only the private costs of inputs but also the costs of the externalities required by their production and use; see *Externality*.

Fungi – one of the five-kingdom classifications; consists of non-photosynthetic, eukaryotic organisms with cell walls, filamentous bodies, and absorptive nutrition.

Fungicide – a chemical that kills *Fungi*.

G

G8 – The Group of Eight is an international forum for the world's major industrialized democracies that emerged following the 1973 oil crisis and subsequent global recession; includes Canada, France, Germany, Italy, Japan, Russia, the U.K., and the U.S., which together represent about 65% of the world economy.

Gaia hypothesis – an ecological hypothesis that proposes that living and nonliving parts of the Earth are a complex interacting system that can be thought of as a single organism; named after the Greek Earth mother goddess Gaia.

Gamma rays – very short wavelength forms of the electromagnetic spectrum.

Gap analysis – a biogeographical technique of mapping biological diversity and endemic species to find gaps between protected areas that leave endangered habitats vulnerable to disruption.

Garden city – a new town designed with special emphasis on landscaping and rural ambiance.

Garden organics – organics derived from garden sources (e.g., prunings, grass clippings).

Gasohol – a mixture of gasoline and ethyl alcohol; used in the internal combustion engine.

Gasoline – a volatile flammable liquid made from petroleum and used as fuel in internal combustion engines; made up of hydrocarbons that are made of 8 carbon chains.

Gene – a unit of heredity; a segment of DNA nucleus of the cell that contains information for the synthesis of a specific protein, such as an enzyme.

Gene banks – storage for seed varieties for future breeding experiments.

Gene Pool – the collective genetic information contained within a population of sexually reproducing organisms.

General fertility rate – the crude birthrate multiplied by the percentage of women of reproductive age.

Generalist species – those able to thrive in a wide variety of environmental conditions and make use of a variety of different resources.

Genetic assimilation – the disappearance of a species as its genes are diluted through crossbreeding with a closely related species.

Genetic diversity – one of the three levels of biodiversity that refers to the total number of genetic characteristics.

Genetic engineering – 1) the use of various experimental techniques to produce molecules of DNA containing new genes or novel combinations of genes, usually for insertion into a host cell for cloning; 2) the technology of preparing recombinant DNA in vitro by cutting up DNA molecules and splicing together fragments from more than one organism; 3) the modification by man of genetic material that would otherwise be subject to the forces of nature only; 4) laboratory manipulation of genetic material using molecular biology techniques to create desired characteristics in organisms.

Genome – the total genetic composition of an organism.

Geometric growth – growth that follows a geometric pattern of increase, such as 2, 4, 8, 16, etc.; see *Exponential growth*.

Geosphere – the solid part of planet Earth, the main divisions being the crust, the mantle and the liquid core; the *Lithosphere* is the part of the geosphere that consists of the crust and the upper mantle.

Geothermal energy – energy derived from the natural heat of the Earth either through geysers, fumaroles, hot springs, or other natural geothermal features or through deep wells that pump heated groundwater; used to generate electricity after transformation.

Germplasm – genetic material that may be preserved for future agricultural, commercial and ecological values (plant seeds or parts or animal eggs, sperm, and embryos).

Global acres – acres/hectares that have been adjusted according to world average biomass productivity so that they can be compared meaningfully across regions; 1 global acre is 1 acre of biologically productive space that has average world productivity; see *Global hectares*.

Global dimming – a reduction in the amount of direct solar radiation reaching the surface of the Earth due to light diffusion as a result of air pollution and increasing levels of cloud (a phenomenon of only the last 30–50 years).

Global environmentalism – a concern for, and action to help solve global environmental problems.

Global hectares – acres/hectares that have been adjusted according to world average biomass productivity so that they can be compared meaningfully across regions; 1 global hectare is 1 hectare of biologically productive space with average world productivity; see also *Global acres*.

Global warming – the observable increase in global temperatures considered mainly caused by the human-induced enhanced greenhouse effect trapping the sun's heat in the Earth's atmosphere; the average temperature of the Earth has risen and fallen over periods of millions of years, such as during Ice Ages; however, current concern is that the increase in greenhouse gases generated by humans, particularly carbon dioxide emissions from use of fossil fuels, will contribute to global warming; preferred term now is "global climate change" because changes in average temperatures have effects on other aspects of weather and climate, including amount of rainfall.

Global warming potential – a system of multipliers devised to enable the warming effects of different gases to be compared.

Globalization – 1) the expansion of interactions to a global or worldwide scale.; 2) the increasing interdependence, integration, and interaction among people and organizations from around the world; 3) a mix of economic, social, technological, cultural, and political interrelationships.

Glyphosate – a nonselective herbicide that is particularly effective against perennial weeds; the active ingredient in the herbicide Roundup™.

Governance – the decision-making procedure; who makes decisions, how they are made, and with what information; the structures and processes for collective decision-making involving both governmental and non-governmental actors.

Grasslands – a biome dominated by grasses and associated herbaceous plants.

Green – in sustainability, a word like "eco-" that is frequently used to indicate consideration for the environment (e.g., green plumbers, green purchasing, etc.); sometimes used as a noun (e.g., the Greens).

Green architecture – building design that moves towards self-sufficient sustainability by adopting *Circular metabolism*.

Green design – the designing of objects, services, and buildings to achieve environmental sustainability.

Green manure – a type of cover crop grown primarily to add nutrients and organic matter to the soil.

Green plans – integrated national environmental plans for reducing pollution and resource consumption while achieving sustainable development and environmental restoration.

Green political parties – political organizations promoting environmental protection, participatory democracy, grassroots organization and sustainable development.

Green power – electricity generated from clean, renewable energy sources (such as solar, wind, biomass or hydropower) and supplied through the grid.

Green products and services – products or services that have a lesser or reduced effect on human health and the environment when compared with competing products or services that serve the same purpose; may include, but are not limited to, those which contain recycled content, reduce waste, conserve energy or water, use less packaging and reduce the number of toxins emitted or consumed.

Green purchasing – purchasing goods and services that minimize impacts on the environment and that are socially just.

Green Revolution – the ongoing transformation of agriculture that led in some places to significant increases in agricultural production between the 1940s and 1960s; it usually requires high inputs of water, plant nutrients, and pesticides.

Green Star – a voluntary rating system for buildings for green design, covering nine impact categories; the highest rating is six stars, which equals a world leader in sustainability.

Green waste – (green organic material or green organics, sometimes referred to as "green wealth") plant material discarded as non-putrescible waste, including tree and shrub cuttings and prunings, grass clippings, leaves, natural (untreated) timber waste and weeds (noxious or otherwise).

Greenhouse effect – the insulating effect of atmospheric greenhouse gases (e.g., water vapor, carbon dioxide, methane, etc.) that keeps the Earth's temperature about 60 °F (16 °C) warmer than it would be otherwise; see *Enhanced greenhouse effect*.

Greenhouse gas – any gas that contributes to the *Greenhouse effect*; gaseous constituents of the atmosphere, both natural and from human activity, which absorbs and re-emit infrared radiation (e.g., CO_2, methane, chlorofluorocarbons (CFCs), nitrous oxide); water vapor (H_2O) is the most abundant greenhouse gas; a natural part of the atmosphere and include carbon dioxide (CO_2), methane (CH_4, which persists for 9 to 15 years, with a greenhouse warming potential (GWP) 22 times that of CO_2), nitrous oxide (N_2O persists for 120 years and has a GWP of 310), ozone (O_3), hydrofluorocarbons, perfluorocarbons and sulfur hexafluoride.

Greenlash – dramatic changes in the structure and dynamic behavior of ecosystems.

Greenwashing – companies that portray themselves as environmentally friendly when their business practices do not support this claim; generally, it applies to unjustified use of green marketing and packaging when the company does not consider their product's total ecological footprint.

Greenwater – water replenishing soil moisture, evaporating from soil, plant and other surfaces and transpired by plants; in nature, the global average amount of rainfall becoming greenwater is about 60%; of the greenwater, about 55% falls on forests, 25% on grasslands and about 20% on crops; productivity can be increased by rainwater harvesting, increased infiltration, and runoff collection; cannot be piped, drunk or sold and is therefore generally ignored by water management authorities but it is crucial to plants in both nature and agriculture and needs careful management as an important part of the global water cycle.

Greywater – household wastewater that has not come into contact with toilet waste; includes water from baths, showers, bathrooms, washing machines, laundry, and kitchen sinks; can be reused without purification for some purposes.

Gross domestic product (GDP) – the total economic activity within national boundaries.

Gross national product (GNP) – the total of all goods and services produced in a national economy; *Gross domestic product* (GDP), in contrast, is used to distinguish economic activity within a country from that of off-shore corporations.

Gross primary productivity – total carbon assimilation.

Groundwater – water located beneath the ground surface in soil pore spaces and the fractures of lithologic formation; does not include water or crystallization held by chemical bonds in rocks or moisture in upper soil layers.

Growth – increase in size, weight, power, etc.

Gully erosion – removal of layers of soil, creating channels or ravines too large to be removed by normal tillage operations.

H

Habitat – an ecological or environmental area that is inhabited by a species.

Habitat conservation plans – agreements under which property owners are allowed to harvest resources or develop the land as long as habitats are conserved or replaced in ways that benefit resident endangered or threatened species in the long run; some incidental "taking" or loss of endangered species is generally allowed in such plans.

Hadley cells – circulation patterns of atmospheric convection currents as they sink and rise in several intermediate bands.

Hard waste – household garbage which is not normally accepted into garbage bins (e.g., old stoves, mattresses).

Hard water – water with high mineral content.

Hazardous chemicals – chemicals that are dangerous, including flammables, explosives, irritants, sensitizers, acids and caustics, etc.; many are hazardous in high concentrations but harmless when diluted.

Hazardous waste – any discarded material containing substances known to be toxic, mutagenic, carcinogenic, or teratogenic to humans or other life-forms; ignitable, corrosive, explosive, or highly reactive alone or with other materials.

Health – a state of physical and emotional wellbeing; the absence of disease or ailment.

Heap-leach extraction – a technique for separating gold from extremely low-grade ores; crushed ore is piled in huge heaps and sprayed with a dilute alkaline-cyanide solution, which percolates through the pile to extract the gold, which is separated from the effluent in a processing plant; this process has a high potential for water pollution.

Heat – energy derived from the motion of molecules; a form of energy into which all other forms of energy may be degraded.

Heat capacity – the amount of heat energy that must be added or subtracted to change the temperature of a body; water has a high heat capacity.

Heat of vaporization – the amount of heat energy required to convert water from a liquid to a gas.

Heat tax – a tax imposed on the use of energy supplies.

Heavy metals – mercury, lead, cadmium, and nickel; highly toxic in very small quantities; can be fatal and can accumulate in the environment with long-term cumulative effects in humans.

Hemoglobin – the iron-containing respiratory pigment in red blood cells of vertebrates, consisting of about 6 percent heme and 94 percent globin.

Herbicide – a chemical that kills or inhibits the growth of a plant.

Herbivore – an organism that eats only plants.

Herbivory – predation in which an organism, known as an herbivore, consumes autotrophs such as plants, algae and photosynthesizing bacteria principally.

Heterarchy – a way of organizing that does not include rank (i.e., each element possesses the same level of importance or authority); compare with *Hierarchy*.

Heterotroph (chemoorganotroph) – 1) an organism that requires organic substrates to obtain its carbon for growth and development; 2) an organism that is incapable of synthesizing its food and must feed upon organic compounds produced by other organisms.

Hidden energy – energy within a system that one is not aware of.

Hierarchy – an organization of parts in which control from the top (generally with few parts), proceeds through a series of levels (ranks) to the bottom (generally consisting of many parts). Compare with *Heterarchy*.

High-density polyethylene (HDPE) – a member of the polyethylene family of plastics; used to make products such as milk bottles, pipes, and shopping bags; may be colored or opaque.

High-level wastes – wastes that are highly radioactive.

High-level waste repository – a place where intensely radioactive wastes can be buried and will be kept unexposed to groundwater and earthquakes for tens of thousands of years.

High-quality energy – intense, concentrated and high-temperature energy that is considered high-quality because of its usefulness in carrying out work.

Histogram – a statistical graph of a frequency distribution in which vertical rectangles of different heights are proportionate to corresponding frequencies; used to graph distributions of populations, such as the percentage of the population in distinct age groups.

Holistic science – the study of entire integrated systems, rather than isolated parts; often takes a descriptive or interpretive approach.

Home energy audits – auditing or analyzing the expenditure of energy in a home, including the loss of energy.

Homeostasis – the property of either an open system or a closed system, especially a living organism, that regulates its internal environment to maintain a stable, constant condition.

Homestead Act – a U.S. legislation passed in 1862 allowing any citizen or applicant for citizenship over 21 years old and the head of a family to acquire 160 acres of public land by living on it and cultivating it for five years.

Homoclime – a region with the same patterns of weather as the one under investigation.

Horsepower (hp) – a unit of power in physics, equal to 550 foot-pounds/sec or 745.7 watts.

Horton overland flow – the tendency of water to flow horizontally across land surfaces when rainfall has exceeded infiltration capacity and depression storage capacity.

Host organism – an organism that provides lodging for a parasite.

Hot desert – deserts of the American Southwest and Mexico; characterized by extreme summer heat and cacti.

House energy rating – an assessment of the energy efficiency of a residential house or unit design, using a 5-star scale.

Household metabolism – the passage of food, energy, water, goods, and waste through the household unit in a similar way to the metabolic activity of an organism.

Human ecology – the study of the interactions of humans with the environment.

Human equivalent (He) – the approximate human daily energy requirement of 12,500 kJ or its approximate energy generating capacity at a basal metabolic rate, which is equivalent to about 80 watts or 3.47222 kWh per day; a 100-watt light bulb, therefore, runs at 1.25 He.

Human resources – human wisdom, experience, skill, labor, and enterprise.

Humus – sticky, brown, insoluble residue from the bodies of dead plants and animals; gives soil its structure, coating mineral particles and holding them together; serves as a major source of plant nutrients.

Hurricanes – large cyclonic oceanic storms with heavy rain and winds exceeding 119 km/hr (74 mph).

Hurricane Nor'easter – a hurricane that generates from the northeast and moves southwest.

Hybrid gas-electric motor – automobiles that run on electric power and small gasoline or diesel engine.

Hydro energy – potential and kinetic energy of water used to generate electricity.

Hydrocarbons – an organic chemical compound containing only hydrogen and carbon atoms, arranged in rows or rings or both, and connected by single, double, or triple bonds; constitute a very large group, including alkanes, alkenes, and alkynes (e.g., petroleum, coal, methane).

Hydroelectric power – the electrical power generated using the power of falling water.

Hydrological cycle (water cycle) – the natural cycle of water from evaporation, transpiration in the atmosphere, condensation (rain and snow) and finally flowing back into the ocean (via rivers).

Hydrosphere – all the Earth's water; this would include water found in the sea, streams, lakes, and other bodies of water, the soil, groundwater, and in the air.

Hydroxyl radical (·OH) – the monovalent group ·OH in such compounds as bases and some acids and alcohols; this radical is characteristic of hydroxides, oxygen acids, alcohols, glycols, phenols, and hemiacetals.

Hypothesis – a tentative explanation that accounts for a set of facts and can be tested for further investigation.

I

Igneous rocks – crystalline minerals solidified from molten magma from deep in the Earth's interior (e.g., basalt, rhyolite, andesite, lava, and granite).

Impervious – incapable of being penetrated.

Inaccessible – not available.

Inbreeding depression – in a small population, an accumulation of harmful genetic traits (through random mutations and natural selection) that lowers the viability and reproductive success of enough individuals to affect the whole population.

Incineration – combustion (by chemical oxidation) of waste material to treat or dispose of that waste material.

Incinerator – an apparatus, such as a furnace, for burning waste.

Indicator species – any biological species that defines a trait or characteristic of the environment.

Indicators – quantitative markers for monitoring progress towards desired goals.

Indirect energy – the energy generated in, and accounted for, by the wider economy as a consequence of an agent's actions or demands.

Inductive reasoning – inferring general principles from specific examples.

Industrial agriculture – a form of modern farming that involves industrialized production of livestock, poultry, fish, or crops.

Industrial ecology – (term originated by Harry Zvi Evan, 1973) the observation that nature produces no waste and therefore provides an example of sustainable waste management which industry should emulate; Natural Capitalism espouses industrial ecology as one of its four pillars, along with energy conservation, material conservation and a redefinition of commodity markets and product stewardship in terms of a service economy.

Industrial Revolution – a period in the late 18th and early 19th centuries when major changes in agriculture, manufacturing, and transportation had a profound effect on socioeconomic and cultural conditions.

Industrial timber – trees used for lumber, plywood, veneer, particleboard, chipboard, and paper; also called "*roundwood.*"

Inertial confinement – a nuclear fusion process in which a small pellet of nuclear fuel is bombarded with extremely high-intensity laser light.

Infiltration – the act or process of infiltrating, as of water into a porous substance or of fluid into the cells of an organ or part of the body.

Informal economy – small-scale individual or family businesses in temporary locations outside the control of normal regulatory agencies.

Inherent value – ethical values or rights that exist as an intrinsic or essential characteristic of a particular thing or class of things simply by the fact of their existence.

Inholdings – private lands within public parks, forests or wildlife refuges.

Insecticide – a pesticide used to control insects in all developmental forms.

Insolation – incoming solar radiation.

In-stream – the use of freshwater where it occurs, usually within a river or stream; includes hydroelectricity, recreation, tourism, scientific and cultural uses, ecosystem maintenance and dilution of waste.

Instrumental value – value or worth of objects that satisfy the needs and wants of moral agents; objects that can be used as a means to some desirable end.

Intangible resources – factors that cannot be contained or measured, such as open space, beauty, serenity, wisdom, diversity, and satisfaction.

Integrated pest management (IPM) – an ecologically-based pest-control strategy that relies on natural mortality factors, such as natural enemies, weather, cultural control methods and carefully applied doses of pesticides; the goal of IPM is not to eliminate all pests but to reduce pest populations to acceptable levels; an ecologically based pest control strategy that relies heavily on natural mortality factors and seeks to use control tactics that disrupt these factors as little as possible.

Integrated product life-cycle management – management of all phases of goods and services to be environmentally friendly and sustainable.

Intercropping – the agricultural practice of cultivating two or more crops in the same space at the same time.

Intergenerational equity – the intention to leave the world in the best possible condition for future generations.

Intergovernmental Panel on Climate Change (IPCC) – the IPCC was established in 1988 by the World Meteorological Organization and the UN Environment Program to provide the scientific and technical foundation for the *United Nations Framework Convention on Climate Change* (UNFCCC), primarily through the publication of periodic assessment reports.

Intermittent – any phenomenon that stops and starts at intervals.

Internal costs – the expenses, monetary or otherwise, borne by those who use a resource.

Internal water footprint – the water embodied in goods produced within a country (although these may be subsequently exported); compare with *External water footprint*.

Internalizing costs – planning so that those who reap the benefits of resource use will also bear all the external costs.

Interplanting – the system of planting two or more crops, either mixed or in alternating rows, in the same field; protects the soil and makes more efficient use of the land.

Interpretive science – an explanation based on observation and description of entire objects or systems rather than isolated parts.

Interspecific competition – in a community, competition for resources between members of different species.

Intrinsic value – the value of something independent of its utility.

Introduced species – a non-native species that has been brought to an area either by accident or intentionally; also known as an "alien species" or "exotic species;" an introduced species may prey upon or compete more successfully with one or more species that are native to the community and thereby alter the entire nature of the community.

Ionizing radiation – high-energy electromagnetic radiation or energetic subatomic particles released by nuclear decay.

Ionosphere – the lower part of the thermosphere.

Ions – electrically charged atoms that have gained or lost electrons.

Irrigation – watering of plants, no matter what system is used.

Irrigation index (Ii) – an efficiency indicator showing the degree of match between water supplied and water used; ideal rating = 1; an Ii of 1.5 would mean a 50% oversupply of water.

Irrigation scheduling – watering plants according to their needs.

Irruptive growth – an expansion in population followed by a dramatic decrease in population; see *Malthusian growth*.

Island biogeography – the study of rates of colonization and extinction of species on islands or other isolated areas based on size, shape, and distance from other inhabited regions.

ISO 14001 – The *International Organization for Standardization*; the international standard for companies seeking to certify their environmental management system; first published in 1996, specifying the requirements for an environmental management system in organizations (companies and institutions), with the goal of minimizing harmful effects on the environment and the goal of continual improvement of environmental performance.

Isobars – lines on a weather map connecting points of equal atmospheric pressure; also called "isopiestic."

Isolated – a population that is separated from other populations of the species (as on an island).

Isotopes – forms of a single element that differ in atomic mass due to a different number of neutrons in the nucleus.

J

J curve – (or "J-shaped curve") a growth curve that depicts exponential growth; called a J curve because of its shape; looks like a "J."

Jet stream – a high-speed, meandering wind current, generally moving from a westerly direction at altitudes of 10 to 15 miles and speeds often exceeding 250 miles per hour; similar to oceanic currents in extent and effect on climate.

Joule (J) – the basic unit of energy; the equivalent of 1 watt of power radiated or dissipated for 1 second; natural gas consumption is usually measured in megajoules (MJ), where 1 MJ = 1, 000,000 J; on large accounts it may be measured in gigajoules (GJ), where 1 GJ = 1,000,000,000 J.

K

Karst – an area of irregular limestone in which erosion has produced fissures, sinkholes, underground streams, and caverns.

Kerosene – a colorless, flammable oil distilled from petroleum and used as a fuel for jet engines, heating, cooking, and lighting.

Keystone species – a species that has a disproportionate effect on its environment relative to its abundance, affecting many other organisms in an ecosystem, and that help in determining the types and numbers of various other species in a community.

Kinetic energy – energy contained in a body because of its motion; equal to one half the mass of the body times the square of its speed (e.g., a rock rolling down a hill, the wind blowing through the trees, water flowing over a dam).

Known resources – those that have been located and are not completely mapped but are likely to become economical in the foreseeable future.

Kwashiorkor – a widespread human protein deficiency disease resulting from a starchy diet low in protein and essential amino acids.

Kyoto Protocol – an international agreement adopted in December 1997 at the Kyoto (Japan) Conference; sets binding emission targets for developed countries that would reduce their emissions, on average, 5.2% below their 1990 levels.

L

Lake Effect snow – lake-generated snow squalls form when cold air passing for long distances over the relatively warm waters of a large lake picks up moisture and heat from the lake and then deposit the moisture in the form of snow upon reaching the downwind shore.

Land reform – democratic redistribution of land ownership to recognize the rights of those who work the land to a fair share of the products of their labor.

Land use, land-use change, and forestry (LULUCF) – land use and land-use changes can act either as sinks or as emission sources; it is estimated that approximately one-fifth of global emissions result from LULUCF activities; the Kyoto Protocol allows parties to receive emissions credit for certain LULUCF activities that reduce net emissions.

Land use planning – a branch of public policy that encompasses various disciplines that seek to order and regulate the use of land efficiently and ethically.

Landfill – solid waste disposal in which refuse is buried between layers of soil; a method often used to reclaim low-lying ground; the word is sometimes used as a noun to refer to the waste itself; also called a "dump," a "tip," and, historically, a "midden."

Landfill gas – the gas emissions from biodegrading waste in a landfill, including carbon dioxide (CO_2), methane (CH_4), and small amounts of nitrogen (N) and oxygen (O) with traces of toluene, benzene, and vinyl chloride.

Landfill levy – 1) taxes applied (at differential rates) to municipal, commercial, and industrial wastes and prescribed wastes disposed to licensed landfills; 2) the taxes used to foster the environmentally sustainable use of resources and best practice in waste management.

Landfill prohibition – the banning of a certain material or product type from disposal to landfills; occasionally occurs, for example, where a preferable waste management option is available.

Landscape ecology – the study of the reciprocal effects of spatial patterns on ecological processes (i.e., how landscape history shapes the features of the land and the organisms that inhabit it as well as our reaction to, and interpretation of, the land).

Landslide – the sudden fall of rock and earth from a hill or cliff; often triggered by an earthquake or heavy rain.

La Niña – is a coupled ocean-atmosphere phenomenon that is the colder counterpart of El Niño, as part of the broader El Niño–Southern Oscillation climate pattern. The name La Niña means "the little girl," analogous to El Niño, meaning "the little boy." During a period of La Niña, the sea surface temperature across the equatorial Eastern Central Pacific Ocean is lower than normal by 5.4 to 9 °F (3 to 5 °C). La Niña persists for at least five months and has effects on the weather across the globe,

Latitude – 1) the angular distance north or south of the Earth's equator, measured in degrees along a meridian, as on a map or globe; 2) a region of the Earth considered in relation to its distance from the equator (e.g., the temperate latitudes).

Law of the minimum – the concept that the growth or survival of a population is directly related to the life requirement that is in the least supply and not to a combination of factors.

LD50 – a chemical dose lethal to 50 percent of a test population.

Leachate (waste) – the mixture of water and dissolved solids (possibly toxic) that accumulates as water passes through waste and collects at the bottom of a landfill site.

Leaching – the movement of a chemical from the upper layers of soil into the lower layers or groundwater by being dissolved in water.

Lead – a soft, malleable, ductile, bluish-white, dense metallic element, extracted chiefly from galena and used in containers and pipes for corrosives, solder and type metal, bullets, radiation shielding, paints and antiknock compounds (atomic number 82; atomic weight 207.19; melting point 327.5 °C; boiling point 1,744 °C; specific gravity 11.35; valence 2, 4).

Leaf area index (LAI) – the ratio of photosynthetic leaf area to the ground area covered; often optimized by shifts in leaf angle, a form of solar tracking; the optimal LAI for photosynthesis is 3-5.

Legionnaires Disease – an acute bacterial respiratory illness caused by the gram-negative bacterium *Legionella pneumophila*, a member of the family *Legionellaceae*; the bacteria have been found in water systems and can survive in the air conditioning systems of large buildings; risk factors for infection include smoking, Chronic Obstructive Pulmonary Disease (COPD), renal failure, cancer, diabetes, and alcoholism.

Less-developed countries (LDC) – non-industrialized nations characterized by low per capita income, high birthrates, and death rates, high population growth rates and low levels of technological development.

Leukemia – malignant neoplasm of blood-forming tissues; characterized by abnormal proliferation of leukocytes; one of the four major types of cancer.

Level (scale, context, or framework) – a context, frame of reference, or degree of organization within an integrated system; may or may not be spatially delimited.

Lifecycle (of a product) – all stages of a product's development, from its raw materials, the manufacturing process, through to its consumption and ultimate disposal.

Life-Cycle Assessment (LCA) – an objective process to evaluate the environmental impacts associated with a product, process or activity; measures the material and energy inputs and outputs at each stage of manufacture, use, and disposal of a product; a means of identifying resource use and waste released to the environment and to assess management options.

Life expectancy – the average age that a newborn infant can expect to attain in a particular time and place.

Lifespan – the longest period of life reached by a type of organism.

Life support systems – according to the *International Union for the Conservation of Nature* (IUCN), the biophysical processes "that sustain the productivity, adaptability, and capacity for renewal of lands, waters or the biosphere as a whole."

Lilac water – recycled water that is unsuitable for drinking.

Limiting factors – chemical or physical factors that limit the existence, growth, abundance, or distribution of an organism; see *Tolerance limits*.

Linear low-density polyethylene – a member of the polyolefin family of plastics; a strong and flexible plastic usually used in the film for packaging, bags and industrial products, such as pressure pipes.

Linear metabolism – direct conversion of resources into wastes that are often sent directly to the landfill.

Lipid – a non-polar organic compound that is insoluble in water but soluble in solvents, such as alcohol and ether; includes fats, oils, steroids, phospholipids, and carotenoids.

Liquid metal fast breeder – a nuclear power plant that converts ^{238}U (uranium 238) to ^{239}Pu (plutonium 239); creates more nuclear fuel than it consumes; because of the extreme heat and density of its core, the breeder uses liquid sodium as a coolant.

Lithosphere – the solid outermost shell of a rocky planet; considered ideal for gardening and agricultural uses.

Loam – a soil composed of sand, silt, and clay in relatively constant concentrations (about 40%-40%-20% concentrations, respectively).

Locally existing capacity – the total ecological production that is found within a country's territories, usually expressed in hectares based on average world productivity.

Lobbying – using personal contacts, public pressure or political action to persuade legislators to vote in a particular manner.

Logistic growth – growth rates regulated by internal and external factors that establish an equilibrium with environmental resources; see *S curve*.

Longevity – the length or duration of life.

Love Canal – an area in Niagara Falls, New York, where seepage from buried toxic wastes contaminated local soil and water; in 1968, President Carter relocated almost all the residents of Love Canal; the impetus for the 1980 *Superfund* legislation.

Low-density polyethylene – a member of the polyolefin family of plastics; a flexible material and usually used as a film for packaging or as bags.

Low entropy energy – high-quality energy, or energy that is concentrated and available; electricity is considered the energy carrier with the lowest entropy (i.e., highest quality) as it can be transformed into mechanical energy at efficiency rates well above 90%; in contrast, fossil fuel chemical energy can only be converted into mechanical energy at a typical efficiency rate of 25% (motor vehicles) to 50% (modern power plants); the chemical energy of biomass is of lower quality.

Low-head hydropower – small-scale hydro technology that can extract energy from small headwater dams; causes much less ecological damage.

Low-level wastes – wastes that are not highly radioactive.

Low-quality energy – diffuse, dispersed energy at a low temperature that is difficult to gather and use for productive purposes.

LULUs – an acronym for "Locally Unwanted Land Uses," such as toxic waste dumps, incinerators, smelters, airports, freeways, and other sources of environmental, economic or social degradation.

M

Magma – molten rock that sometimes forms beneath the surface of the Earth (or any other terrestrial planet) that often collects in a magma chamber and is ejected by volcanos; called "lava" when it spews from an erupting volcano.

Magnetic confinement – a technique for enclosing a nuclear fusion reaction in a powerful magnetic field inside a vacuum chamber.

Malignant tumor – a mass of cancerous cells that have left their site of origin, migrated through the body, invade normal tissues, and growing out of control.

Malnourishment – a nutritional imbalance caused by a lack of specific dietary components or an inability to absorb or utilize essential nutrients.

Malthusian growth – a population explosion followed by a population crash; see also *Irruptive growth*.

Man and Biosphere (MAB) program – a design for nature preserves that divides protected areas into zones with different purposes; a highly protected core is surrounded by a buffer zone and peripheral regions in which multiple-use resource harvesting is permitted.

Mantle – a hot, pliable layer of rock that surrounds the Earth's core and underlies the cool, outer crust.

Manure – organic matter used as fertilizer in agriculture.

Marasmus – a widespread human protein deficiency disease caused by a diet low in calories and protein or imbalanced in essential amino acids.

Marginal costs – the cost to produce one additional unit of a good or service.

Marine – living in or pertaining to the sea.

Marine climate – as its name suggests, west coast marine climates (Cfb) are generally found on the western sides of continents in the belt of the westerly winds between roughly 40 to 60 degrees latitude; this location produces a humid climate, often quite rainy, with mild temperatures considering the fairly high latitudes; this is the effect of having large bodies of water to windward; water is a great modifier of temperatures because it heats and cools slowly; the proximity of water to windward leads to much milder winter temperatures and somewhat cooler summer temperatures that are experienced at continental locations at the same latitudes (Cfb climates are considered by some to be gloomy climates because they are the world's cloudiest climates; distinctive kind of biological community adapted to those conditions).

Market benefits – benefits of a climate policy that can be measured in terms of avoided market impacts, such as changes in resource productivity (e.g., lower agricultural yields, scarcer water resources) and damages to human-built environments (e.g., coastal flooding due to sea-level rise).

Market equilibrium – the dynamic balance between supply and demand under a given set of conditions in a "free" market (one where there are no monopolies and no government interventions).

Marsh – wetland without trees; in North America, this type of land is characterized by cattails and rushes.

Mass burn – incineration of unsorted solid waste.

Material flow – the cycling of materials, which is driven by the flow of energy.

Material identification – words, numbers or symbols used to designate the composition of components of a product or packaging (Note: a material identification symbol does not indicate whether an item can be recycled).

Materials recovery facility (MRF) – a center for the reception and transfer of materials recovered from the waste stream; at an MRF, materials are also sorted by type and treated (e.g., cleaned, compressed, etc.).

Matter – anything that takes up space and has mass.

Mauna Loa record – the record of measurement of atmospheric CO_2 concentrations taken at Mauna Loa Observatory, Mauna Loa, Hawaii, since March 1958, showing the continuing increase in average annual atmospheric CO_2 concentrations.

Maximum soil water deficit – the amount of water stored in the soil that is readily available to plants.

Meander – a turning or winding of a stream.

Mediation – an informal dispute resolution process in which parties are encouraged to discuss issues openly but in which all decisions are reached by consensus and any participant can withdraw at any time.

Mediterranean climate areas – specialized landscapes with warm, dry summers, cool, wet winters, many unique plants, and animal adaptations and many levels of endemism.

Megacity – a large city, typically with a population above ten million; see *Megalopolis*.

Megadiverse countries – The 17 countries that are home to the largest fraction of wild species; Australia is one such country.

Megalopolis – indicates an urban area with more than 10 million inhabitants; also known as a "*Megacity*" or "super city."

Megawatt (MW) – unit of electrical power equal to 1,000 kilowatts or 1 million watts.

Mesosphere – the atmospheric layer above the stratosphere and below the thermosphere; the middle layer; temperatures are usually very low.

Metabolism – all the energy and matter exchanges that occur within a living cell or organism; collectively, the life processes.

Metamorphic rock – igneous and sedimentary rocks modified by heat, pressure and chemical reactions.

Methane hydrate – small bubbles or individual molecules of methane (natural gas) trapped in a crystalline matrix of frozen water.

Micro-hydro generators – small power generators that can be used in low-level rivers to provide economic power for four to six homes, freeing them from dependence on large utilities and foreign energy supplies.

Microorganism – an organism visible only through a microscope (e.g., a virus, bacterium or fungus).

Microwave – a high-frequency electromagnetic wave, one millimeter to one meter in wavelength, intermediate between the infrared and short-wave radio wavelengths.

Middle East – a region in western Asia and northeast Africa including 15 countries: Bahrain, Islamic Republic of Iran, Iraq, Israel, Jordan, Kuwait, Lebanon, Oman, Qatar, Saudi Arabia, Syria, the United Arab Emirates, and Yemen.

Migration – the moving of one species or a group of species from one area to another.

Milankovitch cycles – periodic variations in the tilt, eccentricity, and wobble of the Earth's orbit; Milutin Milankovitch suggested that they are responsible for cyclic weather changes.

Milpa agriculture – an ancient farming system in which small patches of tropical forests are cleared and perennial polyculture (growing many crops in the same space) agriculture is practiced, which is then followed by many years of fallow to restore the soil; see *Swidden agriculture*.

Mineral – a naturally occurring, inorganic, crystalline solid with definite chemical composition and characteristic physical properties.

Mitigation – repairing or rehabilitating a damaged ecosystem or compensating for damage by providing a substitute or replacement area.

Mixed perennial polyculture – growing a mixture of different perennial crop species (where the same plant persists for more than one year) together in the same plot.

Mobile garbage bin – a wheeled curbside container for the collection of garbage or other materials.

Moderators – a substance, for example, graphite or beryllium, that slows neutrons in a nuclear reactor so that they can bring about the fission of uranium.

Molecule – a combination of two or more atoms.

Monitored, retrievable storage – holding wastes in underground mines or secure surface facilities such as dry casks where they can be watched and repackaged, if necessary.

Monkeywrenching – environmental sabotage, such as driving large spikes in trees to protect them from loggers, vandalizing construction equipment, pulling up survey stakes for unwanted developments, and destroying billboards; see *Ecotage*.

Mono Lake – an oasis in the dry Great Basin in California and a vital habitat for millions of migrating and nesting birds.

Monoculture – the practice of producing or growing one single crop over a wide area.

Monoculture agroforestry – intensive planting of a single species; an efficient wood production approach, but one that encourages pests and disease infestations and conflicts with wildlife habitat or recreation uses.

Monsoon – 1) a wind system that influences large climatic regions and reverses direction seasonally; 2) a wind from the southwest or south that brings heavy rainfall to southern Asia in the summer and/or the rain that accompanies this wind.

Montane coniferous forests – coniferous forests of the mountains consisting of belts of different forest communities along an altitudinal gradient.

Montreal Protocol – a treaty, signed in 1987, that governs stratospheric ozone protection and research and the production and use of ozone-depleting substances; provides for ending production of ozone-depleting substances such as CFCs; under the Montreal Protocol, various research groups continue to assess the ozone layer; the Multilateral Fund provides resources to developing nations to promote their transition to ozone-safe technologies.

Moral agents – beings capable of making distinctions between right and wrong and acting accordingly (i.e., those who are held responsible for their actions).

Moral extensionism – expansion of our understanding of inherent value or rights to persons, organisms, or things that might not be considered worthy of value or rights under some ethical philosophies.

Moral subjects – beings that are not capable of distinguishing between right or wrong or that are not able to act on moral principles and yet are capable of being wronged by others.

Morals – a set of ethical principles that guide our actions and relationships.

Morbidity – describes how often a disease afflicts a certain area (e.g., "how many people have lung cancer in New York City").

More-developed countries (MDC) – industrialized nations characterized by high per capita incomes, low birth and death rates, low population growth rates, and high levels of industrialization and urbanization.

Mortality – 1) the death rate in a population; 2) the probability of dying.

Mortality rate – understood as the total number of deaths per 1,000 people of a given age group.

Mullerian mimicry – the evolution of two noxious species who share common predators to resemble each other; the theory is that predators, who associate the unique characteristics of one noxious species with danger, will also choose to ignore the mimicking species.

Mulch – any composted or non-composted organic material, excluding plastic that is suitable for placing on soil surfaces to restrict moisture loss from the soil and to provide a source of nutrients to the soil.

Multiple use – many uses that occur simultaneously; used in forest management; limited to mutually compatible uses.

Municipal sewage – the wastewater from households, offices and other buildings in a city; can either be sanitary sewage only or sanitary sewage and stormwater; collected at treatment plants where solids are removed by "primary sewage treatment" and then treated by various other methods including using aerobic bacteria to remove organic wastes in "secondary treatment" and advanced or "tertiary treatment" with various chemical and physical processes.

Municipal waste – solid waste generated from domestic premises (garbage and hard waste) and local government activities such as street sweeping, litter and street tree trimming; also includes waste dropped at transfer stations and construction waste from owner/occupier renovations.

Mutagens – agents, such as chemicals or radiation, that damage or alter genetic material (DNA) in cells.

Mutation – a change, either spontaneous or by external factors, in the genetic material of a cell; mutations in the gametes (egg or sperm cells) can be inherited by future generations of organisms.

Mutualism – a symbiotic relationship between individuals of two different species in which both species benefit from the association.

N

NAAQS – "*National Ambient Air Quality Standard*;" federal standards specifying the maximum allowable levels (averaged over specific periods) for regulated pollutants in *Ambient air*.

Natality – the production of new individuals by birth, hatching, germination or cloning.

National ambient air quality standards – health-based pollutant concentration limits established by the EPA that apply to the outside air.

National Packaging Covenant – a self-regulatory agreement between packaging industries and government.

National Priority List (NPL) – set up by the EPA as part of the *Superfund* program; locates and sets priorities for cleaning up hazardous waste sites.

Native species – species that are originally found in a certain area.

Natural – the existing air, water, land and energy resources from which all resources derive; main functions include resource production (such as fish, timber or cereals), waste assimilation (such as CO_2 absorption, sewage decomposition) and life-support services (ultraviolet radiation, biodiversity, water cleansing, climate stability); the environmental services that must be maintained so that human development can be sustainable.

Natural capital – natural resources and ecological processes that are equivalent to financial capital.

Natural history – the study of where and how organisms carry out their lifecycles.

Natural increase – the crude death rate subtracted from the crude birth rate.

Natural resources – naturally occurring substances that are considered valuable in their relatively unmodified (natural) form.

Natural selection – the process by which favorable heritable traits become more common in successive generations of a population of reproducing organisms, and the more unfavorable heritable traits become less common.

Neighborhood environment improvement plan – plans developed by a local community including residents, special interest groups, local government, local industry, and government agencies.

Nematocide – a chemical that kills nematodes (e.g., roundworms and threadworms).

Neo-classical economics – a branch of economics that attempts to apply the principles of modern science to economic analysis in a mathematically rigorous, non-contextual, abstract, predictive manner.

Neo-Luddites – people who reject technology, believing it is the cause of environmental degradation and social disruption; named after the followers of Ned Ludd, who tried to turn back the Industrial Revolution in England by wrecking factories.

Neo-Malthusian – a belief that the world is characterized by scarcity and competition in which too many people fight over too few resources; named for Thomas Malthus, who predicted a dismal cycle of misery, vice, and starvation as a result of human overpopulation.

Net energy yield – total useful energy produced during the lifetime of an entire energy system minus the energy used, lost, or wasted in making useful energy available.

Net primary production – the energy or biomass content of plant material that has accumulated in an ecosystem over a period of time through photosynthesis; the amount of energy left after subtracting the respiration of primary producers (mostly plants) from the total amount of solar energy that is fixed biologically; gross primary productivity minus respiratory losses (this is the carbon gain).

Neurotoxins – toxic substances, such as lead or mercury, that specifically poison nerve cells.

Neutron – a subatomic particle, found in the nucleus of the atom, that has no electromagnetic charge.

New towns – experimental urban environments that seek to combine the best features of the rural village and the modern city.

Niches – a specific category that an organism fits into in an environment and their role in carrying out the processes in that ecosystem.

Nickel-cadmium batteries – batteries typically used in appliances such as power tools and mobile phones; cadmium is a heavy metal that poses a risk to human and ecosystem health.

Nihilists – those who reject all moral or religious beliefs, often concluding that the world and everything in it amounts to nothingness ("nihil" is Latin for "nothing") or is meaningless.

NIMBY – an acronym for "Not In My BackYard;" the rallying cry of those opposed to *LULUs*.

Nitrates – 1) salt or ester of nitric acid; 2) sodium nitrate or potassium nitrate used as a fertilizer.

Nitrate-forming bacteria – bacteria that combine ammonia with oxygen to form nitrites that can be used by green plants to build proteins.

Nitrogen cycle – the circulation and utilization of nitrogen in nature, consisting of a cycle of chemical reactions in which atmospheric nitrogen is compounded, dissolved in the rain and deposited in the soil, where it is assimilated and metabolized by bacteria and plants, eventually returning to the atmosphere by bacterial decomposition of organic matter; the specific principles include: N_2 is the most abundant gas in atmosphere (78%), nitrogen-fixing bacteria convert it to NH_3, and nitrate-forming bacteria combine NH_3 with oxygen to form NO_2 and then NO_3, plants absorb and make NH_4, consumers eat plants, nitrogen re-enters the environment when these organisms die, shed, urinate, produce excrement, which de-nitrifying bacteria break down into N_2 and the process repeats.

Nitrogen-fixing bacteria – bacteria that convert nitrogen from the atmosphere or soil solution into ammonia that can then be converted to plant nutrients by nitrite- and nitrate-forming bacteria.

Nitrogen oxides (NOx) – highly reactive gases formed when nitrogen in fuel or combustion air is heated to over 650 °C (1,200 °F) in the presence of oxygen or when bacteria in soil or water oxidize nitrogen-containing compounds; often mentioned in discussions of nitrogen-based air pollution as a reference to both nitric oxide (NO) and nitrogen dioxide (NO_2); in addition to particulates and sulfur dioxide, NOx is one of the major pollutants related to energy use; can transform to nitrates in the atmosphere.

Noise pollution (environmental noise) – displeasing human-created or machine-created sound that disrupts the activity or happiness of human or animal life.

Non-criteria pollutants – all pollutants not specifically mentioned in NAAQS or the *Hazardous Air Pollutants* (HAPs) in the Clean Air Act (e.g., benzene, dioxins, pesticides); see *Unconventional air pollutants*.

Nongovernmental organizations (NGOs) – a term referring collectively to pressure and research groups, advisory agencies, political parties, professional societies and other groups concerned about environmental quality, resource use, and many other issues.

Non-ferrous metals – those metals that contain little or no iron (e.g., copper, brass and bronze).

Nonpoint source pollution – water pollution affecting a water body from diffuse sources, rather than a point source which discharges to a water body at a single location.

Nonpoint sources – scattered, diffuse sources of pollutants, such as runoff from farm fields, golf courses, construction sites, etc.

Nonrenewable resources – minerals, fossil fuels and other materials present in essentially finite amounts (within human time scales) in our environment.

Nor'easter – a storm blowing from the northeast.

North/South division – a description of the fact that most of the world's wealthier countries tend to be in North America, Europe, and Japan, while the poorer countries tend to be located closer to the equator.

No-till farming – considered a kind of conservation tillage system and is sometimes called "zero tillage."

Nuclear energy – energy released by reactions within atomic nuclei, as in nuclear fission or fusion; also called "atomic energy."

Nuclear fission – the radioactive decay process in which isotopes split apart to create two smaller atoms.

Nuclear fusion – a process in which two smaller atomic nuclei fuse into one larger nucleus and release energy; the source of power in a hydrogen bomb.

Nucleic acids – large organic molecules made of nucleotides that function in the transmission of hereditary traits, in protein synthesis and control of cellular activities.

Nucleus – 1) the center of the atom; occupied by protons and neutrons; 2) in cells, the organelle that contains the chromosomes (DNA).

Nuées ardentes – deadly, denser-than-air mixtures of hot gases and ash ejected from volcanoes.

Numbers pyramid – a diagram showing the relative population sizes at each trophic level in an ecosystem; usually corresponds to the biomass pyramid.

Nutrient – chemicals required for the growth of organisms; phosphorus, nitrogen, and potassium are major plant nutrients, but there are also many trace elements, needed in small quantities for the growing and developing of animal and plant life.

O

Ocean acidification – the reduction in pH of oceans caused by their uptake of anthropogenic carbon dioxide from the atmosphere)

Ocean shorelines – rocky coasts and sandy beaches along the oceans; support rich, stratified communities.

Ocean thermal electric conversion (OTEC) – energy derived from the temperature differential between warm ocean surface waters and cold, deep waters; this differential can be used to drive turbines attached to electric generators.

Oceania – the islands of the southern, western, and central Pacific Ocean, including Melanesia, Micronesia, and Polynesia; in some definitions, Oceania extends to encompass Australia, New Zealand, and Maritime Southeast Asia.

Oceanic islands – islands in the ocean; formed by breaking away from a continental landmass, volcanic action, coral formation, or a combination of sources; support distinctive communities.

Offset – something that balances, counteracts, or compensates.

Offset allowances – a controversial component of air quality regulations that allows a polluter to avoid installation of control equipment on one source by an "offsetting" pollution reduction at another source.

Ogallala aquifer – the largest aquifer in North America, located under the Great Plains in the U.S.

Oil glut – when the supply of oil on the market greatly exceeds demand, resulting in lower oil prices.

Oil shale – a fine-grained sedimentary rock that is rich in solid organic material called "kerogen;" when heated, the kerogen liquefies to produce a fluid petroleum fuel.

Old-growth forests – forests free from disturbance for long enough (generally 150 to 200 years) to have mature trees, ideal physical conditions, species diversity, and other characteristics of equilibrium ecosystems; see *Ancient forest*.

Oligotrophic – the condition of rivers and lakes that have clear water and low biological productivity (oligo = little; trophic = nutrition); usually clear, cold, infertile headwater lakes and streams.

Omnivore – a species of animal that eats both plants and animals as primary food sources.

OPEC – acronym for "*Organization of Petroleum Exporting Countries;*" founded in 1960 to unify and coordinate petroleum policies of the members.

Open access system – a commonly held resource for which there are no management rules.

Open burning – uncontrolled fires in an open dump.

Open canopy – a forest where tree crowns cover less than 20 percent of the ground; see also *Woodland*.

Open-pit mining – a method of extracting rock or minerals from the earth by their removal from an open pit or burrow; also called "*opencast mining*" or "*open-cut mining.*"

Open range – unfenced, natural grazing lands; includes woodland as well as grassland.

Open system – a system that exchanges energy and matter with its environment.

Operational energy – the energy used in carrying out a particular operation.

Optimum – the most favorable condition regarding an environmental factor.

Orbital – the space or path in which an electron orbits the nucleus of an atom.

Organic – derived from a living organism; containing carbon.

Organic agriculture – a holistic production management system that avoids the use of synthetic fertilizers, pesticides, and genetically modified organisms (GMO), minimizes pollution of air, soil, and water and optimizes the health and productivity of interdependent communities of plants, animals, and people.

Organic compounds – complex molecules organized around skeletons of carbon atoms arranged in rings or chains; includes biomolecules, molecules synthesized by living organisms.

Organic gardening – gardening that follows, in general principle, the philosophy of *Organic agriculture*.

Organic matter – compounds that contain carbon and hydrogen covalently bonded together in molecules; molecules from living matter; organic wastes in sewage and runoff from lawns and farms in freshwaters can cause oxygen-depletion and degradation of water quality.

Organics – plant or animal matter originating from domestic or industrial sources (e.g., grass clippings, tree pruning, food waste).

Orographic effect (Chinook winds) – a moist wind blowing from the sea on the Northwestern U.S. coast.

Overburden – overlying layers of noncommercial sediments that must be removed to reach a mineral or coal deposit.

Overdrawn – taking too much out or depleting resources (e.g., pumping water from an aquifer at a faster rate than it can be replenished or recharged by rainfall).

Overnutrition – occurs when the amount of nutrients ingested exceeds what is needed for normal growth, metabolism, and development; also called "hyperalimentation."

Overshoot – growth beyond an area's carrying capacity; ecological deficit occurs when human consumption and waste production exceed the capacity of the Earth to create new resources and absorb waste; during overshoot, natural capital is being liquidated to support current use, so the Earth's ability to support future life declines.

Oxidation – the act or process of oxidizing (i.e., to change a compound by increasing the proportion of the electronegative part or charge (an element or ion) from a lower to a higher positive valence); removing one or more electrons from an atom, ion or molecule to combine with oxygen.

Oxygen cycle – the circulation and utilization of oxygen in the biosphere.

Oxygen sag – oxygen decline downstream from a pollution source that introduces materials with high biological oxygen demands.

Ozone – a colorless, highly reactive molecule containing three oxygen atoms; a dangerous pollutant in ambient air that is soluble in alkalis and cold water; in the stratosphere, ozone forms an ultraviolet absorbing shield that protects us from mutagenic radiation; a strong oxidizing agent and can be produced by electric discharge in oxygen or by the action of ultraviolet radiation on oxygen in the stratosphere.

P

Pacific Decadal Oscillation (PDO) – a large pool of warm water that moves north and south in the Pacific Ocean every 30 years or so and has large effects on North America's climate.

Parabolic mirrors – curved mirrors that focus light from a large area onto a single central point, thereby concentrating solar energy and producing high temperatures.

Paradigm – a model that provides a framework for interpreting observations.

Parasite – an organism that lives in or on another organism, deriving nourishment at the expense of its host, usually without killing it.

Parent – an original radioactive atom or any material.

Parsimony – the reluctance to use resources or spend money.

Particulate material – atmospheric aerosols, such as dust, ash, soot, lint, smoke, pollen, spores, algal cells, and other suspended materials; originally applied only to solid particles but now extended to droplets of liquid.

Parts per billion (ppb) – the number of parts of a chemical found in 1 billion parts of a gas, liquid, or solid mixture.

Parts per million (ppm) – the number of parts of a chemical found in 1 million parts of a gas, liquid, or solid mixture.

Parts per trillion (ppt) – the number of parts of a chemical found in 1 trillion parts of a gas, liquid, or solid mixture.

Passive heat absorption – the use of natural materials or absorptive structures without moving parts to gather and hold heat; the simplest and oldest use of solar energy.

Patchiness – within a larger ecosystem, the presence of smaller areas that differ in some physical conditions and support somewhat different communities; a diversity-promoting phenomenon.

Pathogen – an organism that produces disease in a host organism, the disease being an alteration of one or more metabolic functions in response to the presence of the organism.

Pathogenic – describes any microorganism capable of causing disease.

Pay-by-weight systems – financial approaches to managing waste that charge prices according to the quantity of waste collected, rather than a price per pick-up or fixed annual charge as is typically applied to households for curbside services; may provide an incentive to reduce waste generation.

Peat – deposits of moist, acidic, semi-decayed organic matter.

Pellagra – a disease characterized by lassitude, torpor, dermatitis, diarrhea, and sometimes dementia and death; brought about by a diet deficient in tryptophan and niacin.

Peptides – two or more amino acids linked by a peptide bond.

Per capita consumption – the average amount of commodity used per person.

Percolation – water slowly moving through soil and gravel into an aquifer.

Perennial species – plants that grow for more than two years.

Permafrost – a permanently frozen layer of soil that underlies the arctic tundra.

Permanent retrievable storage – placing waste storage containers in a secure building, salt mine, or bedrock cavern where they can be inspected periodically and retrieved, if necessary.

Persistent organic pollutants (POPs) – chemical compounds that persist in the environment and remain biologically active for many years.

Pervious surface – one which can be penetrated by air and water.

Pest – any organism that reduces the availability, quality or value of a useful resource.

Pest resurgence – the rebound of pest populations due to acquired resistance to chemicals and nonspecific destruction of their natural predators and competitors by broad-scale pesticides.

Pesticide – any substance or mixture of substances intended for preventing, destroying or controlling any pest; includes substances intended for use as a plant growth regulator, defoliant, desiccant or agent for thinning fruit or preventing the premature fall of fruit and substances applied to crops either before or after harvest to protect the commodity from deterioration during storage and transport.

Pesticide treadmill – a need for constantly increasing doses of pesticide or new, stronger pesticides to prevent pest resurgence.

Petrochemicals – chemicals synthesized from oil.

pH – a value that indicates the acidity or alkalinity of a solution on a scale of 0 to 14, based on the proportion of H+ ions present.

pH scale – p(**p**otential of) H(**h**ydrogen); the logarithm of the reciprocal of hydrogen-ion concentration in gram atoms per liter; used as a measure of the acidity or alkalinity of a solution on a scale of 0 to 14 (where 7 is neutral).

Phosphates – 1) a salt or ester of a phosphoric acid; 2) the trivalent anion PO_{43}, derived from phosphoric acid H_3PO_4, an organic compound of phosphoric acid in which the acid group is bound to nitrogen or a carboxyl group in a way that permits useful energy to be released (as in metabolism); 3) a phosphatic material used for fertilizers.

Phosphorous (phosphorus) cycle – the movement of phosphorus atoms from rocks and soil through the biosphere and hydrosphere and back to the soil.

Photochemical oxidants – products of secondary atmospheric reactions; see *Smog*.

Photodegradable plastics – plastics that break down when exposed to sunlight or a specific wavelength of light.

Photosynthesis – the transformation of radiant energy to chemical energy by plants; the manufacture by plants of carbohydrates from carbon dioxide and water; driven by energy from sunlight, catalyzed by chlorophyll, releasing oxygen as a byproduct; the capture of the sun's energy (primary production) to power all life on Earth (consumption).

Photosynthetic efficiency – the percentage of available sunlight captured by plants and used to make useful products.

Photovoltaic – the direct conversion of light into electricity.

Photovoltaic cell – an energy-conversion device that captures solar energy and directly converts it to electrical current.

Physical or abiotic factors – nonliving factors, such as temperature, light, water, minerals, and climate that influence an organism.

Phytoplankton – microscopic, free-floating, autotrophic organisms that function as producers in aquatic ecosystems; see *Autotroph* and *Plankton*.

Pioneer species – in primary succession on a terrestrial site, the plants, lichens, and microbes that first colonize the site.

Plague – a disease that spreads very rapidly, infecting very large numbers of people and killing a great many of them; an outbreak of such a disease.

Plankton – mostly microscopic animal and plant life suspended in water; a valuable food source for fish and other marine animals; see *Phytoplankton*.

Plant quality – a standard of plant appearance or yield.

Plasma – a hot, electrically neutral gas of ions and free electrons.

Plastic – one of many high-polymeric substances, including both natural and synthetic products but excluding rubbers; at some stage in its manufacture, every plastic is capable of flowing, under heat, and pressure, if necessary, into a mold of the desired final shape.

PM-10 – particulates that are less than 10 microns in diameter; present in the smoke created by burning wood.

Poachers – those who hunt wildlife illegally.

Point sources – specific locations of highly concentrated pollution discharge, such as factories, power plants, sewage treatment plants, underground coal mines, and oil wells.

Policy – a societal plan or statement of intentions intended to accomplish some social good.

Policy cycle – the process by which problems are identified and acted upon in the public arena.

Political economy – the branch of economics concerned with modes of production, distribution of benefits, social institutions, and class relationships.

Polluter Pays Principle (PPP) – the principle that producers of pollution should, in some way, compensate others for the effects of their pollution.

Pollution – to make foul, unclean, or dirty; any physical, chemical or biological change that adversely affects the health, survival, or normal activities of living organisms or that alters the environment in undesirable ways.

Pollution charges – fees assessed per unit of pollution based on the *Polluter pays principle* (PPP).

Polycentric complex – cities with several urban cores surrounding a once-dominant central core.

Polyethylene terephthalate (PET) – a clear, tough, light, and shatterproof type of plastic, used to make products such as soft drink bottles, film packaging, and fabrics.

Polypropylene (PP) – a member of the polyolefin family of plastics; light, rigid, and glossy and is used to make products such as washing machine agitators, clear film packaging, carpet fibers, and housewares.

Polystyrene (PS) – a member of the styrene family of plastics; easy to mold and is used to make refrigerator and washing machine components; can be foamed to make single-use packaging, such as cups, meat and produce trays.

Polyvinyl chloride (PVC) – a member of the vinyl family of plastics; can be clear, flexible, or rigid and is used to make products such as fruit juice bottles, credit cards, pipes, and hoses.

Population – a group of individuals of the same species occupying a given area.

Population crash – a sudden population decline caused by predation, waste accumulation, or resource depletion; see *Dieback*.

Population dieback – when the growth of a population slows due to some factor.

Population explosion – the growth of a population at exponential rates to a size that exceeds environmental carrying capacity; usually followed by a *Population crash*.

Population momentum – a potential for increased population growth as young members reach reproductive age.

Pore spaces – the amount of space available for groundwater due to the topography of the area.

Porosity – the ratio of the volume of all the pores in a material to the volume of the whole.

Post-consumer material or waste – a material or product that has served its intended purpose and has been discarded for disposal or recovery; includes returns of material from the distribution chain, waste that is collected and sorted after use and curbside waste; compare with *Pre-consumer waste*.

Post-materialist values – a philosophy that emphasizes the quality of life over the acquisition of material goods.

Post-modernism – a philosophy that rejects and often mocks the optimism and universal claims of modern positivism.

Potable – water that is safe to drink.

Potential energy – the energy of a particle or system of particles derived from its position or condition, rather than motion (e.g., a raised weight, a coiled spring or a charged battery).

Power – the rate at which work is done; electrically, Power = Current x Voltage or "P = I V."

Precautionary principle – where there are threats of serious irreversible environmental damage, lack of full scientific certainty should not be used as a reason for not introducing measures to prevent that degradation (*Rio Declaration*); see *UNFCCC*.

Precedent – an act or decision that can be used as an example in dealing with subsequent similar situations.

Precipitation – in weather, any liquid or solid water particles that fall from the atmosphere to the Earth's surface; includes drizzle, rain, snow, snow pellets, ice crystals, ice pellets, and hail.

Precipitator – pollution control device that collects particles from an air stream.

Pre-consumer material or waste – material diverted to the waste stream during a manufacturing process; waste from manufacture and production.

Precycling – making environmentally sound decisions at the store and reducing waste.

Predation – the act of feeding by a *Predator*.

Predator – an organism that feeds directly on other organisms to survive; live-feeders, such as herbivores and carnivores.

Pre-industrial – referring to the time before industrialization (i.e., before the Industrial Revolution c. 1750-1850).

Prescribed waste and prescribed industrial waste – those wastes listed in the Environment Protection (Prescribed Waste) Regulations of 1998 and subject to requirements under the industrial waste management policy of 2000; carry special handling, storage, transport and often licensing requirements and attract substantially higher disposal levies than non-prescribed solid wastes.

Prevention of significant deterioration – a clause of the Clean Air Act that prevents degradation of existing clean air; opposed by the industry as an unnecessary barrier to development.

Price elasticity – a situation in which the supply of and the demand for a commodity will fluctuate with price changes.

Primary energy – forms of energy obtained directly from nature; the energy in raw fuels (electricity from the grid is not primary energy), used mostly in energy statistics when compiling energy balances.

Primary pollutants – chemicals released directly into the air in a harmful form.

Primary producers – producers that are responsible for a substantial amount of the food for the rest of the food chain in an ecosystem.

Primary productivity – a synthesis of organic materials (biomass) by green plants using the energy captured in photosynthesis.

Primary (sewage) treatment – a process that removes solids from sewage before it is discharged or treated further.

Primary standards – regulations of the 1970 Clean Air Act; intended to protect human health.

Primary succession – an ecological succession that begins in an area where no biotic community previously existed.

Principle of competitive exclusion – a result of natural selection whereby two similar species in a community occupy different ecological niches, thereby reducing competition for food.

Producer – in ecology, an organism that synthesizes food molecules from inorganic compounds by using an external energy source; most producers are photosynthetic.

Producer responsibility – the legal responsibilities of producers/manufacturers for the full life of their products.

Product – 1) a thing produced by labor; 2) the material items bought in shops; 3) in ecology, the results of photosynthesis.

Product stewardship – the principle of shared responsibility by all sectors involved in the manufacture, distribution, use, and disposal of products for the consequences of these activities; the manufacturer's responsibility extending to the entire life of the product.

Production frontier – the maximum output of two competing commodities at different levels of production.

Productivity (ecology) – the rate at which radiant energy is used by producers to form organic substances as food for consumers.

Prokaryotic – cells that do not have a membrane-bounded nucleus or membrane-bounded organelles.

Promoters – agents that are not carcinogenic but that assist in the progression and spread of tumors; sometimes called "co-carcinogens."

Pronatalist pressures – influences that encourage people to have children.

Proteins – chains of amino acids linked by peptide bonds.

Proton – a positively charged subatomic particle found in the nucleus of an atom.

Proven resources – those resources that have been thoroughly mapped and are economical to recover at current prices with available technology.

Provisioning services – one of the major ecosystem services: the products obtained from ecosystems (e.g., genetic resources, food, fiber, and freshwater).

Proximity – the state, quality, sense or fact of being near or next to; closeness.

Public trust – a doctrine obligating the government to maintain public lands in a natural state as guardians of the public interest.

Pull factors – in urbanization, conditions that attract people from the country to the city.

Push factors – in urbanization, conditions that force people out of the country and into the city.

Pyrolysis – advanced thermal technology involving the thermal decomposition of organic compounds in the complete absence of oxygen, under pressure and at an elevated temperature.

Q

Qualitative – of or concerning a trait, characteristic, or property.

Quantitative – relating to or expressed as a specified or indefinite number or amount.

R

Radiative forcing – changes in the energy balance of the earth-atmosphere system in response to a change in factors such as greenhouse gases, land-use change or solar radiation; positive radiative forcing increases the temperature of the lower atmosphere, which in turn increases temperatures at the Earth's surface; negative radiative forcing cools the lower atmosphere; most commonly measured in units of watts per square meter or "W/m^2."

Radiatively active gases – a gas that occurs naturally or is produced anthropogenically that affects atmospheric radiation by absorption or emission.

Radioactive – an unstable isotope that decays spontaneously and releases subatomic particles or units of energy.

Radioactive decay – a change in the nuclei of radioactive isotopes that spontaneously emit high-energy electromagnetic radiation or subatomic particles while gradually changing into another isotope or a different element.

Radionuclides – isotopes that exhibit radioactive decay.

Radon – a radioactive gaseous element formed by the disintegration of radium; the heaviest of the inert gasses; occurs naturally (especially in areas over granite) and is considered a hazard to health.

Rain garden – an engineered area for the collection, infiltration and *Evapotranspiration* of rainwater runoff, mostly from impervious surfaces; reduces rain runoff by allowing stormwater to soak into the ground (as opposed to flowing into storm drains and surface waters which can cause erosion, water pollution, flooding and diminished groundwater; can also absorb water contaminants that would otherwise end up in water bodies; the term arose in Maryland in the 1990s as a more marketable expression for *Bioremediation*.

Rain shadow – a dry area on the downwind side of a mountain.

Rainforest – a forest with high humidity, constant temperature and abundant rainfall (generally over 150 inches per year); can be tropical or temperate.

Rainwater harvesting – collecting rainwater either in storages or the soil mostly close to where it falls; the attempt to increase rainwater productivity by storing it in ponds, wetlands, etc. and helping to avoid the need for infrastructure to bring water from elsewhere; when practiced on a large scale upstream, it reduces the available water downstream.

Rangeland – grasslands and open woodlands suitable for livestock grazing.

Rational choice – public decision-making based on reason, logic, and science-based management.

Raw materials – materials that are extracted from the ground and processed (e.g., bauxite is processed into aluminum).

Reasonably Available Control Technology (RACT) – the lowest emissions limit that a source can meet by the application of control technology that is reasonably available considering technological and economic feasibility.

Recharge zones – an area where water filters into aquifers.

Reclaimed water – water is taken from a waste (effluent) stream and purified to a level suitable for further use.

Reclamation – chemical, biological or physical cleanup and reconstruction of severely contaminated or degraded sites to return them to something like their original topography and vegetation.

Recoverable resources – those accessible with current technology but deemed not to be economical under current conditions.

Recovered material – (waste) material that would have otherwise been disposed of as waste or used for energy recovery but has instead been collected and recovered (reclaimed) as material input, thus avoiding the use of new primary materials.

Recovery rate – (waste) the recovery rate is the percentage of materials consumed that is recovered for recycling.

Re-creation – construction of an entirely new biological community to replace one that has been destroyed on that or another site.

Recreational fishing – by the 1890s, most states in the U.S. had put restrictions on fishing; today, a fishing license is needed to fish for recreation in lakes and inland bodies of water.

Recyclables – strictly, all materials that may be recycled, but this may include the recyclable containers and paper/cardboard component of curbside waste (excluding garden organics).

Recycled content – proportion, by mass, of recycled material in a product or packaging; only pre-consumer and post-consumer materials are considered as recycled content.

Recycled material – waste that has been converted into usable forms, not necessarily in its original use; see *Recovered material*.

Recycled water – treated stormwater, *Greywater,* or *Blackwater* suitable for uses like toilet flushing, irrigation, industry, etc.; non-drinking water, so indicated using a lilac non-drinking label.

Recycling – reprocessing of discarded materials into new, useful products; not the same as *Reuse* of materials for their original purpose, but the terms are often used interchangeably.

Red tide – a population explosion or bloom of minute, single-celled marine organisms called "dinoflagellates;" billions of these cells can accumulate in protected bays where the toxins they contain can poison other marine life.

Reduced tillage systems – systems, such as minimum-till, conserve-till, and no-till, that preserve soil, save energy and water, and increase crop yields.

Reflected – to return light rays from a surface in such a way that the angle at which a given ray is returned is equal to the angle at which it strikes the surface.

Reforestation – 1) replanting of forests on lands that have recently been harvested; 2) the direct human conversion of non-forested land to forested land through planting, seeding or promotion of natural seed sources on land that was once forested but now no longer so; according to the language of the *Kyoto Protocol*, for the first commitment period from 2008–2012, reforestation activities are limited to reforestation occurring on lands that did not contain forest at the beginning of 1990.

Reformer – a device that strips hydrogen from fuels such as natural gas, methanol, ammonia, gasoline or vegetable oil so they can be used in a fuel cell.

Refracted – to alter the course of a wave of energy that passes into something from another medium, as water does to light entering it from the air; caused by differences in wave speed.

Refuse-derived fuel – the processing of solid waste to remove metal, glass, and other unburnable materials; the organic residue is shredded, formed into pellets, and dried to make fuel for power plants.

Regenerative farming – farming techniques and land stewardship that restore the health and productivity of the soil by rotating crops, planting ground cover, protecting the surface with crop residue and reducing synthetic chemical inputs and mechanical compaction.

Regional consequences – the impact of global climate change varies from one region to another; some dry areas may become wetter; another region may have less precipitation.

Regulating services – (sustainability) the benefits obtained from the regulation of ecosystem processes including, for example, the regulation of climate, water, or disease.

Regulations – rules established by administrative agencies; can be more important than *Statutory law* in the day-to-day management of resources.

Rehabilitate land – a utilitarian program to make an area useful to humans.

Rehabilitation – to rebuild elements of structure or function in an ecological system without necessarily achieving complete restoration to its original condition.

Relative – the relation of one thing to another; expressed as the ratio of the specified quantity to the total magnitude (as the value of a measured quantity) or the mean of all the quantities involved.

Relative humidity – at any given temperature, a comparison of the actual water content of the air with the amount of water that could be held at saturation.

Relativists – those who believe moral principles are always dependent on the situation.

REM – acronym for "*roentgen equivalent man*;" a unit used in radiation protection to measure the amount of damage to human tissue from a dose of ionizing radiation; the amount of ionizing radiation required to produce the same biological effect as one "*rad*" of high-penetration x-rays; an average American receives about 0.370 rems of radiation per year.

Remediation – cleaning up chemical contaminants from a polluted area.

Renewable energy – any source of energy that can be used without depleting its reserves; these sources include solar energy and other sources like wind, wave, biomass, geothermal, and hydro energy.

Renewable energy certificates – Market trading mechanisms created through the Renewable Energy (Electricity) Act 2000 in connection with the Canadian government's mandatory renewable energy target; provide a "premium" revenue stream for electricity generated from renewable sources.

Renewable resources – resources normally replaced or replenished by natural processes; resources not depleted by moderate use (e.g., solar energy, biological resources such as forests and fisheries, biological organisms, and some biogeochemical cycles).

Renewable water supplies – annual freshwater surface runoff plus annual infiltration into underground freshwater aquifers that are accessible for human use.

Reprocessing (waste) – changing the physical structure and properties of a waste material that would otherwise have been sent to landfill, to add financial value to the processed material; this may involve a range of technologies including composting, anaerobic digestion and energy from waste technologies such as pyrolysis, gasification, and incineration.

Reservoir – a natural or artificial pond or lake used for the storage and regulation of water.

Residence time – the length of time a component, such as an individual water molecule, will spend in a compartment or location before it moves on through a process or cycle.

Residual waste – waste that remains after the separation of recyclable materials (including *Green waste*).

Residue – 1) what is leftover or remains; residues of some contaminants may remain after clean-up; 2) the part of a molecule that remains after a portion of its constituents are removed.

Resilience – the ability of a community or ecosystem to recover from disturbances.

Resistant – the ability of an individual or community to resist being changed by potentially disruptive events.

Resource – in economic terms, anything with potential use in creating wealth or giving satisfaction.

Resource Conservation and Recovery Act (RCRA) – regulates the handling of wastes from the cradle to the grave; establishes rules for the handling of such waste from the time it is generated, while it is being packaged, stored, while it is transported and how it is disposed of, as well as the disposal sites themselves.

Resource flow – the totality of changes in multiple resource stocks, or at least any pair of them, over a specified period.

Resource intensity – ratio of resource consumption relative to its economic or physical output (e.g., liters of water used per dollar spent or liters of water used per ton of aluminum produced); at the national level, energy intensity is the ratio of total primary energy consumption of the country to either the gross domestic product, or the ratio of physical output to total goods produced.

Resource productivity – the output obtained for a given resource input.

Resource partitioning – in a biological community, various populations sharing environmental resources through specialization, thereby reducing direct competition; see *Ecological niche.*

Resource recovery – (waste) the process of obtaining matter or energy from discarded materials.

Resource scarcity – a shortage or deficit in some *Resource*.

Resource stock – the total amount of a resource often related to resource flow (the number of resources harvested or used per unit of time); to harvest a resource stock sustainably, the harvest must not exceed the net production of the stock; measured in mass, volume, or energy and flows in mass, volume or energy per unit of time.

Respiration – (biology) 1) uptake by a living organism of oxygen from the air (or water) which is then used to oxidize organic matter or food; the outputs of this oxidation are usually CO_2 and H_2O; 2) the metabolic process by which organisms meet their internal energy needs and release CO_2.

Restoration – to bring something back to a former condition; ecological restoration involves active manipulation of nature to recreate the conditions that existed before human disturbance.

Restoration ecology – seeks to repair or reconstruct ecosystems damaged by human actions.

Retail therapy – using shopping to obtain a "lift" to make up for other things lacking in our lives.

Retrofit – to replace an existing item with an updated version.

Reuse – the second pillar of the waste hierarchy; recovering value from a discarded resource without reprocessing or remanufacture (e.g., clothes sold through second-hand shops strictly represent a form of re-use, rather than *Recycling)*.

Reverse osmosis – a process of desalinization where water is forced under pressure through a semipermeable membrane whose tiny pores allow water to pass but exclude most salts and minerals.

Riders – amendments attached to bills in conference committee, often completely unrelated to the bill to which they are added.

Rill erosion – the removing of thin layers of soil caused by little rivulets of running water cutting small channels in the soil.

Risk – the probability that something undesirable will happen as a consequence of exposure to a hazard.

Risk assessment – evaluation of the short-term and long-term risks associated with a activity or hazard; usually compared to anticipated benefits in a cost-benefit analysis.

RNA – Ribonucleic acid; nucleic acid used for transcription and translation of the genetic code found on DNA molecules.

Rock – a solid, cohesive aggregate of one or more crystalline minerals.

Rock cycle – the process whereby rocks are broken down by chemical and physical forces; sediments are moved by wind, water and gravity, settle and reform into rock, and then eventually are crushed, folded, melted and re-crystallized into new forms.

Routinely monitored – regular, periodic testing.

Ruminant animals – cud-chewing animals, such as cattle, sheep, goats, and buffalo, with multi-chambered stomachs in which cellulose is digested with the aid of bacteria.

Runoff – the excess of precipitation over evaporation; water that can't be absorbed by the ground; the main source of surface water and, in broad terms, the water available for human use.

Run-of-the-river flow – ordinary river flow not accelerated by dams, flumes, etc.; some small, modern, high-efficiency turbines can generate useful power using only run-of-the-river flow, with a current of only a few kilometers per hour.

Rural area – an area in which most residents depend on agriculture or the harvesting of natural resources for their livelihood.

S

S curve – a curve that depicts logistic growth; called an S curve because of its shape; also called an "S-shaped curve;" see *Logistic growth*.

Saffir/Simpson – a scale to measure hurricanes based on wind speeds and air pressure.

Salinity – (ecology) dissolved salts in water and soils, generally in the context of human activity such as clearing and planting for annual crops rather than perennial trees and shrubs; can make soils infertile.

Salinization – a process in which mineral salts accumulate in the soil, killing plants; occurs when soils in dry climates are irrigated profusely.

Salt domes – a solid mass of salt that was once fluid but has flowed into fractures in surrounding rock and geologic structures.

Saltwater intrusion – the movement of saltwater into freshwater aquifers in coastal areas where groundwater is withdrawn faster than it is replenished.

Sanitary landfills – a landfill in which refuse and municipal waste are buried every day under enough soil or fill to eliminate odors, vermin, and litter.

Saturation point – the maximum concentration of water vapors the air can hold at a given temperature.

Scale – the physical dimensions, in either space or time, of phenomena or events.

Scattered – few and far apart in distance or time.

Scavenger – an organism that feeds on the dead bodies of other organisms.

Scientific method – a systematic, precise, objective study of a problem; generally, this requires observation, hypothesis development and testing, data gathering, and interpretation.

Scientific theory – an explanation supported by many tests that have come to be accepted by a consensus of scientists.

Scrubbers – an air pollution device that uses a spray of water or reactant, or a dry process to trap pollutants in emissions.

Second law of thermodynamics – states that with each successive energy transfer or transformation in a system, less energy is available to do work.

Secondary energy – primary energies that are transformed in energy conversion processes to more convenient secondary forms, such as electrical energy or cleaner fuels.

Secondary pollutants – chemicals modified to a hazardous form after entering the air or that are formed by chemical reactions as components of the air mix and interact.

Secondary recovery technique – pumping pressurized gas, steam, or chemical-containing water into a well to squeeze more oil from a reservoir.

Secondary standards – regulations of the 1972 Clean Air Act intended to protect materials, crops, visibility, climate, and personal comfort.

Secondary succession – succession on a site where an existing community has been disrupted.

Secondary treatment – bacterial decomposition of suspended particulates and dissolved organic compounds that remain after primary sewage treatment.

Sectors – (economics) economic groupings used to generalize patterns of expenditure and use.

Secure landfill – a solid waste disposal site lined and capped with an impermeable barrier that prevents leakage or *Leaching*; drain tiles, sampling wells, and vent systems provide monitoring and pollution control.

Sediment – (ecology) soil or other particles that settle to the bottom of water bodies.

Sedimentary rock – deposited material that remains in place long enough or has been covered with enough material to compact into stone; examples include shale, sandstone, breccia, and conglomerates.

Sedimentation – the deposition of organic materials or minerals by chemical, physical or biological processes.

Seismic activity – describes the size, type, and frequency of earthquakes in an area over time.

Selective cutting – harvesting only mature trees of certain species and size; usually more expensive than clear-cutting, but it is less disruptive for wildlife and often better for forest regeneration.

Self-organization – the process by which systems use energy to develop structure and organization.

Self-regulating – an internal mechanism by which a system or organism controls its functions.

Sentinel indicator – (ecology) an indicator that captures the essence of the process of change affecting a broad area of interest and which is also easily communicated.

Septic sewage – sewage in which anaerobic respiration is taking place; characterized by a blackish color and the smell of hydrogen sulfide.

Septic tank – a type of sedimentation tank in which the sludge is retained long enough for the organic content to undergo anaerobic digestion; typically used for receiving the sewage from houses and other premises that are too isolated for connection to a sewer.

Sequestration – (global warming) the removal of carbon dioxide from the Earth's atmosphere and storing it in a natural sink, as when trees absorb CO_2 in photosynthesis and store it in their tissues.

Seriously undernourished – those who receive less than 80 percent of their minimum daily caloric requirements.

Sewage – water and raw effluent disposed of through toilets, kitchens, and bathrooms; includes water-borne wastes from domestic uses of water from households or similar uses in trade or industry.

Sewer – a pipe conveying sewage.

Sewerage – a system of pipes and mechanical appliances for the collection and transportation of domestic and industrial sewage.

Sewerage system – sewage system infrastructure; the network of pipes, pumping stations and treatment plants used to collect, transport, treat and discharge sewage.

Sewer-mining – tapping directly into a sewer (either before or after a sewage treatment plant) and extracting wastewater for treatment and use.

Shallow ecology – a critical term applied to superficial environmentalists who claim to be green but are quick to compromise and who do little to bring about fundamental change.

Shantytowns – settlements created when people move onto undeveloped lands and build their shelter with cheap or discarded materials; some are illegal subdivisions where a landowner rents land without city approval, others are land invasions.

Sheet erosion – peeling off thin layers of soil from the land surface; accomplished primarily by wind and water.

Shredder flock – the residue from shredded car bodies, white goods (i.e., large home appliances) and the like.

Sick Building Syndrome – a building whose occupants experience acute health and/or comfort effects that appear to be linked to the time spent there, but where no specific illness or cause can be identified; complaints may be localized in a particular room or zone, or may spread throughout the building.

Siltation – to become choked or obstructed with silt or mud.

Simple living – a lifestyle that individuals may pursue a variety of motivations, such as spirituality, health or ecology; others may choose simple living for reasons of social justice or a rejection of consumerism; some may do so out of an explicit rejection of "westernized values," while others choose to live more simply for reasons of personal taste, a sense of fairness or for personal economy; distinguished from the simple lifestyles of those living in conditions of poverty in that its proponents are consciously choosing to not focus on lifestyle directly tied to money or cash-based economics; see *Voluntary simplicity*.

Sinkholes – a large surface crater caused by the collapse of an underground channel or cavern; often triggered by groundwater withdrawal.

Sinks – 1) processes or places that remove or store gases, solutes, or solids; 2) any process, activity, or mechanism that results in the net removal of greenhouse gases, aerosols, or precursors of greenhouse gases from the atmosphere.

Slow Food – the Slow Food movement was founded in Italy in 1986 by Carlo Petrini as a response to the negative impact of multinational food industries; a counteracting force to Fast Food as it encourages using local, seasonal produce, restoring time-honored methods of production and preparation and sharing food at communal tables; encourages environmentally sustainable production, ethical treatment of animals and social justice; gatherings of Slow Food supporters are called "*convivial*;" Slow Food enthusiasts seek to defend biodiversity in our food supply, to better their lives by appreciating the sensation of taste and to celebrate the connection between plate and planet.

Sludge – a semi-solid mixture of organic and inorganic materials that settle out of wastewater at a sewage treatment plant.

Slums – legal but inadequate multifamily tenements or rooming houses; some are custom built for rent to poor people; others have been converted from some other use.

Smart growth – efficient use of land resources and existing urban infrastructure.

Smog – 1) (photochemical) air pollution produced by the action of sunlight on hydrocarbons, nitrogen oxides, and other pollutants; 2) (industrial) primarily a winter phenomenon that occurs when sulfur dioxide emissions and smoke particles react with water vapor; 3) the term used to describe the combination of industrial smoke (and automobile exhaust) and fog formerly in the stagnant air of London and in the air of present-day Los Angeles.

Social ecology – a socialist/humanist philosophy based on the communitarian anarchism of the Russian geographer Peter Kropotkin; shares much with *Deep ecology* except that it is more humanist in its outlook.

Social justice – equitable access to resources and the benefits derived from them; a system that recognizes people's inalienable rights and adheres to what is fair, honest, and moral.

Sodicity – (ecology) a measure of the sodium content of soil; sodic soils are dispersible and vulnerable to erosion.

Sodification – the build-up in soils of sodium relative to potassium and magnesium in the composition of the exchangeable cations (positively charged ions) of the clay fraction.

Soil – a complex mixture of weathered mineral materials from rocks, partially decomposed organic molecules, and a host of living organisms.

Soil acidification – reduction in pH, usually in the soil; can result in poorly structured or hard-setting topsoil that cannot support enough vegetation to prevent erosion.

Soil bulk density – the relative density of soil measured by dividing the dry weight of soil by its volume.

Soil compaction – the degree of compression of soil; heavy compaction can impede plant growth.

Soil conditioner – any composted or non-composted material of organic origin that is produced or distributed for adding to soils; includes "soil amendment," "soil additive," "soil improver" and similar materials but excludes polymers that do not biodegrade (e.g., plastics, rubbers, coatings).

Soil horizons – horizontal layers that can be analyzed to reveal a soil's history, characteristics, and usefulness.

Soil moisture deficit – the volume of water needed to raise the soil water content of the root zone to "field capacity," which is the amount of water held by soil after the excess has been drained.

Soil organic carbon (SOC) – the total organic carbon of a soil exclusive of carbon from the undecayed plant and animal residue.

Soil organic matter (SOM) – the organic fraction of the soil exclusive of undecayed plant and animal residues.

Soil structure – the way soil particles are collected into aggregates or "crumbs;" important for the passage of air and water.

Soil water storage – the total amount of water stored in the soil in the plant root zone.

Solar energy – the radiant energy of the sun, which can be converted into other forms of energy, such as heat or electricity.

Solar power – electricity generated from solar radiation.

Solid industrial waste – solid waste generated from commercial, industrial, or trade activities, including waste from factories, offices, schools, universities, state and federal government operations, and commercial construction and demolition work; excludes wastes that are prescribed under the *Environment Protection Act* of 1970 and quarantine wastes.

Solid inert waste – hard waste and dry vegetative material which has a negligible activity or effect on the environment, such as demolition material, concrete, bricks, plastic, glass, metals, and shredded tires.

Solid waste – non-hazardous, non-prescribed solid waste materials ranging from municipal garbage to industrial waste; generally, domestic and municipal, commercial and industrial, construction and demolition, etc.

Soluble – susceptible to being dissolved in a liquid, particularly in water.

Source separation – (waste) separation of recyclable material from other waste at the point where and at the time when the waste is generated (i.e., at its source); includes separation of recyclable material into its component categories, e.g., paper, glass, aluminum; may include further separation within each category (e.g. paper into computer paper, office whites and newsprint); the practice of separating materials into discrete materials streams prior to collection by or delivery to reprocessing facilities.

Southern pine forest – United States coniferous forest ecosystem characterized by a warm, moist climate.

Specialist species – those that can only thrive in a narrow range of environmental conditions and/or have a limited diet.

Species – a taxonomic category subordinate to a genus (or subgenus) and superior to a subspecies or variety, composed of individuals possessing common characters and that can reproduce sexually among themselves but cannot produce fertile offspring when mated with other organisms, distinguishing them from other categories of individuals of the same taxonomic level; in taxonomic nomenclature, species are designated by the genus name followed by a Latin or Latinized adjective or noun.

Species diversity – the number and relative abundance of species present in a community.

Species recovery plan – a plan for restoration of an endangered species through protection, habitat management, captive breeding, disease control or other techniques that increase populations and encourage survival.

Specific heat capacity – the amount of energy needed to increase the temperature of 1 kg of a substance by 1 °C; can be considered a measure of resistance to an increase in temperature and important for energy saving.

Spectrum – an ordered array of the components of an emission or wave.

Spent fuel – the uranium cores that are taken out of the nuclear power plant.

Spillways – a passage permitting surplus water to run over or around an obstruction (such as a dam).

Spontaneous – 1) happening or arising without apparent external cause; self-generated; 2) arising from a natural inclination or impulse and not from any external incitement or constraint; 3) unconstrained and unstudied in manner or behavior; 4) plants growing without cultivation or human labor; indigenous.

Sport hunting – hunting animals, not just for food.

Sprawl – an unlimited outward extension of city boundaries that lowers population density, consumes open space, generates freeway congestion, and causes decay in central cities.

Spring overturn – a springtime lake phenomenon that occurs when the surface ice melts and the surface water temperature warms to its greatest density at 4 °C and then sinks, creating a convection current that displaces the nutrient-rich bottom waters.

Squatter towns – *Shantytowns* that occupy land without the owner's permission; some are highly organized movements in defiance of authorities; others grow gradually.

Stability – 1) in ecological terms, a dynamic equilibrium among the physical and biological factors in an ecosystem or a community; 2) relative homeostasis.

Stable runoff – the fraction of water available year-round; usually more important than total runoff when determining human uses.

Stack emissions – emissions of pollutants from a smokestack, including dioxin, furans, nitrogen oxides, and carbon.

Stakeholders – parties having an interest in a project or outcome.

Standard Metropolitan Statistical Area (SMSA) – an urbanized region with at least 100,000 inhabitants with strong economic and social ties to a central city of at least 50,000 people.

Standing – the right to take part in legal proceedings.

State Environment Protection Policies – statutory instruments under the Environment Protection Act of 1970 that identify beneficial uses of the environment that are to be protected, establish environmental indicators and objectives and set forth attainment programs to implement the policies.

State of the environment reporting – a scientific assessment of environmental conditions, focusing on the impacts of human activities, their significance for the environment, and social responses to the identified trends.

Stationary energy – that energy that is other than transport fuels and fugitive emissions; used mostly for the production of electricity but also for manufacturing, processing and in agriculture, fisheries, etc.

Statute law – formal documents or decrees enacted by the legislative branch of government.

Statutory law – rules passed by a state or national legislature.

Steady-state – a constant pattern (e.g., a balance of inflows and outflows).

Steady-state economy – characterized by low birth and death rates, use of renewable energy sources, recycling of materials, and emphasis on durability, efficiency, and stability.

Sterilization – 1) making an organism barren or infertile (unable to reproduce); 2) to clear an object of living organisms by heating it or by application of chemicals.

Stewardship – a philosophy that holds that humans have a unique responsibility to manage, care for, and improve nature.

Storm surge – giant waves, often fifty miles wide and twenty-five feet or more high, that are caused by the force of a hurricane. (As the eye of the hurricane makes landfall, the wave comes sweeping across the coastline; aided by the hammering effect of the breaking of the waves, it acts like a giant bulldozer sweeping everything in its path).

Stormwater – rainfall that accumulates in natural or artificial systems after heavy rain; surface run-off or water sent to drains during heavy rain.

Strategic Environmental Assessment (SEA) – a system of incorporating environmental considerations into policies, plans, and programs; especially in the European Union.

Strategic Lawsuits Against Public Participation (SLAPP) – lawsuits that have no merit but are brought merely to intimidate and harass private citizens who act in the public interest.

Strategic metals and minerals – materials a country cannot produce itself but that it needs to use for essential materials or processes.

Stratosphere – the zone in the atmosphere extending from the tropopause to about 30 miles above the Earth's surface; temperatures are stable or rise slightly with altitude; has very little water vapor but is rich in ozone; above the troposphere and below the mesosphere.

Strip cutting – harvesting trees in strips narrow enough to minimize edge effects and to allow natural regeneration of the forest.

Strip farming – planting different kinds of crops in alternating strips along land contours; when one crop is harvested, the other crop remains in place to protect the soil and prevent water from running straight down a hill.

Strip mining – removing surface layers over coal seams using giant earth-moving equipment; creates a huge open pit from which enormous surface-operated machines scoop coal and transported by trucks; an alternative to deep mines.

Structure – in ecological terms, patterns of organization, both spatial and functional, in a community.

Sublimation – the process by which water can move between solid and gaseous states without ever becoming liquid.

Subsidence – a settling of the ground surface caused by the collapse of porous formations that result from the withdrawal of large amounts of groundwater, oil, or other underground materials.

Subsoil – a layer of soil beneath the topsoil that has lower organic content and higher concentrations of fine mineral particles; often contains soluble compounds and clay particles that have been carried down by percolating water.

Sulfur cycle – the chemical and physical reactions by which sulfur moves into or out of storage and through the environment.

Sulfur dioxide – a colorless, corrosive gas directly damaging to both plants and animals.

Sulfur oxides – a molecule formed by the combination of sulfur and oxygen (SOx).

Sullage – domestic wastewater from baths, basins, showers, laundries, kitchens and floor waste (but not from toilets).

Superfund – a fund established by Congress in 1984 to pay for containment, cleanup or remediation of abandoned toxic waste sites; financed by fees paid by toxic waste generators and by cost-recovery from cleanup projects.

Supply – the quantity of a product being offered for sale at various prices, other things being equal.

Supporting services – (sustainability) ecosystem services that are necessary to produce all other ecosystem services (e.g., biomass production, production of atmospheric oxygen, soil formation, nutrient and water cycling).

Surface mining – when minerals are mined from surface pits; see *Strip mining*.

Surface runoff – the part of rainfall passing out of an area into the drainage system.

Surface tension – a condition in which the water surface meets the air and acts as an elastic skin.

Survivorship – the percentage of a population reaching a given age or the proportion of the maximum lifespan of the species reached by any individual.

Suspended particulate matter (SPM) (aerosols) – a suspension or dispersion of fine particles of a solid or liquid in a gas.

Suspended solids (SS) – solid particles suspended in water; used as an indicator of water quality.

Sustainability – the Brundtland Commission Report's definition is: "Sustainable development is a development that meets the needs of the present without compromising the ability of future generations to meet their own needs."

Sustainability covenant – Under Section 49 of the Environment Protection Act of 1970, a sustainability covenant is an agreement which a person or body undertakes to increase the resource use efficiency and/or reduce environmental impacts of activities, products, services and production processes; parties can voluntarily enter into such agreements with the EPA, or could be required to if they are declared by a Governor in Council (Canada), on the recommendation of the EPA or have the potential for a significant impact on the environment.

Sustainability science – the multidisciplinary scientific study of sustainability, focusing especially on the quantitative dynamic interactions between nature and society; its objective is a deeper and more fundamental understanding of the rapidly growing interdependence of the nature-society system and the intention to make this sustainable; critically examines the tools used by sustainability accounting and the methods of sustainability governance.

Sustainability triangle – a graphic indication of the action needed to stabilize CO_2 levels below about 500 ppm; shows different stabilization "wedges," indicating the savings made per year using a particular strategy.

Sustainable agriculture – an ecologically sound, economically viable, socially just, and humane agricultural system; stewardship, soil conservation, and integrated pest management are essential for sustainability.

Sustainable consumption – sustainable resource use, a change to society's historical patterns of consumption and behavior that enables consumers to satisfy their needs with better-performing products or services that use fewer resources, cause less pollution and contribute to social progress worldwide.

Sustainable development – a real increase in the wellbeing and standard of life for the average person that can be maintained over the long-term without degrading the environment or undermining the ability of future generations to meet their own needs.

Sustained yield – utilization of a renewable resource at a rate that does not impair or damage its ability to be fully renewed on a long-term basis.

Swale – an open channel transporting surface run-off to a drainage system, usually grassed; promotes infiltration, the filtration of sediment by plants, and often has an ornamental interest.

Swamp – a wetland with trees, such as the extensive swamp forests of the southeastern United States, especially Florida and Louisiana.

Swidden agriculture – land which has been cleared for cultivation, left to regenerate and repeated; see *Milpa agriculture*.

Symbiosis – the intimate living together of members of two different species; includes mutualism, commensalism, and, in some classifications, parasitism.

Synergistic effects – when an injury caused by exposure to two environmental factors is together greater than the sum of exposure to each factor separately would be.

Synfuels – synthetic gas or synthetic oil made from coal or other sources.

System – a set of parts organized into a whole, usually processing flow of energy; a group of interacting, interrelated or interdependent elements forming a complex whole.

Systemic – a condition or process that affects the whole body; many metabolic poisons are systemic.

T

Taiga – the northernmost edge of the boreal forest, including species-poor woodland and peat deposits; intergrading with the arctic tundra.

Tailings – mining waste left after mechanical or chemical separation of minerals from crushed ore.

Take-back – a concept commonly associated with product stewardship, placing responsibility on brand-owners, retailers, manufacturers, or other supply chain partners to accept products returned by consumers once they have reached the end of their useful life; products may then be recycled, treated or sent to landfill.

Taking – unconstitutional confiscation of private property.

Tar sands – sand deposits containing petroleum or tar.

Tax incentive – reduction in taxes given to encourage a specific behavior on the part of the recipient.

Technological optimists – those who believe that technology and human enterprise will find cures for all problems; also called "*Promethean environmentalism.*"

Technosphere – synthetic and composite components and materials formed by human activity; true technosphere materials, like plastics, are not biodegradable.

Tectonic plates – huge blocks of the Earth's crust that slide around slowly, pulling apart to open new ocean basins or crashing ponderously into each other to create new, larger landmasses.

Temperate – with moderate temperatures, weather, or climate; neither hot nor cold; a mean annual temperature between 0 and 20 °C.

Temperate rainforest – the cool, dense, rain forest of the northern Pacific coast; shrouded in fog much of the time; dominated by large conifers many hundreds of years old.

Temperature – a measure of the speed of motion of a typical atom or molecule in a substance.

Temperature inversions – an atmospheric condition in which a layer of warm air traps cooler air near the Earth's surface, preventing the normal rising of surface air.

Tennessee Valley Authority – (TVA), a federal corporation, created by the Congress in 1933 to operate Wilson Dam and to develop the Tennessee River and its tributaries in the interest of navigation, flood control and the production and distribution of electricity; enactments include reforestation, industrial and community development, test-demonstration farming, the development of fertilizer and the establishment of recreational facilities; includes a number of dams for electricity and flood control.

Teratogens – chemicals or other factors that specifically cause abnormalities during embryonic growth and development.

Terracing – shaping the land to create level shelves of Earth to hold water and soil; requires extensive hand labor or expensive machinery, but it enables farmers to farm very steep hillsides.

Territoriality – an intense form of intraspecific competition in which organisms define as theirs an area surrounding their home site or nesting site and defend it, primarily against other members of their species.

Tertiary treatment – the removal of inorganic minerals and plant nutrients after primary and secondary treatment of sewage.

Thermal mass – (architecture) any mass that can absorb and store heat and can, therefore, be used to buffer temperature change; concrete, bricks, and tiles need much heat energy to change their temperature and therefore have high thermal mass; timber has low thermal mass.

Thermal plume – a plume of hot water discharged into a stream or lake by a heat source, such as a power plant.

Thermal pollution – industrial discharge of heated water into a river, lake or another body of water, causing a rise in temperature that endangers aquatic life.

Thermocline – in oceans and lakes, a distinctive temperature transition zone that separates an upper layer that is mixed by the wind (the epilimnion) and a colder, deeper layer that is not mixed (the hypolimnion).

Thermodynamics – a branch of physics that deals with transfers and conversions of energy; *First law of thermodynamics*: energy can be transformed and transferred but cannot be destroyed or created; *Second law of thermodynamics*: with each successive energy transfer or transformation, less energy is available to do work.

Thermosphere – the highest atmospheric zone; a region of hot, dilute gases above the mesosphere extending out to about 1,000 miles from the Earth's surface.

Third pipe system – a third pipe, in addition to the standard water supply pipe and sewer disposal pipe, which carries recycled water for irrigation purposes.

Third World – less-developed countries that are neither capitalistic and industrialized (the First World) nor centrally-planned socialist economies (the Second World).

Threatened species – while it may still be abundant in parts of its territorial range, a species that has declined significantly in total numbers and may be on the verge of extinction in certain regions or localities.

Three Gorges Dam – near Yichang on the Yangtze River in China; helps control the flooding of the Yangtze River Valley; in addition, multiple river flows make the Three Gorges complex the largest electricity-generating facility in the world; a lake about 400 miles long was formed behind the dam, forcing the relocation of more than a million people and permanently submerging many historical sites.

Threshold – (ecology) a point that, when crossed, can bring rapid and sometimes unpredictable change in a trend (e.g., the sudden altering of ocean currents and the sea level due to the increasingly rapid melting of ice at the poles).

Tidal/ocean/wave energy – mechanical energy from water movement used to generate electricity.

Tidal station – a dam built across a narrow bay or estuary that traps tidewater flowing both in and out of the bay; water flowing through the dam spins turbines attached to electric generators.

Timberline – in the mountains, the highest-altitude edge of the forest that marks the beginning of the treeless alpine tundra.

Tipping fee – a fee for disposal of waste.

Tolerance limits – maximum and minimum requirements in which an individual or system can maintain normal processes; see *Limiting factors*.

Topography – a detailed map of the contours of surfaces of land.

Topsoil – usually the uppermost 3 to 10 inches of soil; however, its thickness ranges from a meter or more under virgin prairie to zero in some deserts; layer in which organic material is mixed with mineral particles; critical for agriculture.

Tornado – a rotating column of air usually accompanied by a funnel-shaped downward extension of a cumulonimbus cloud and having a vortex several hundred yards in diameter whirling destructively at speeds of up to 500 miles per hour.

Tort law – civil court cases that seek compensation for damages.

Total energy use – the total of combined direct and indirect energy use.

Total fertility rate – the number of children that, on average, a woman is likely to have in her lifetime at present age-specific fertility rates; calculated as the average number of children born to women of each given age in a particular year; an average for women of all childbearing ages for that year is derived from this.

Total growth rate – the net rate of population growth resulting from births, deaths, immigration, and emigration.

Total maximum daily loads (TMDL) – the amount of a pollutant that a water body can receive from both point and non-point sources and still meet water quality standards.

Total water use – in water accounting, distributed water use + self-extracted water use + reuse water use; used to mean total direct and indirect water use.

Town water – water supplied by government or private enterprise and known as the mains or reticulated water supply.

Toxic – poisonous; a substance that reacts with specific cellular components to kill cells.

Toxic colonialism – shipping toxic wastes to a weaker or poorer nation.

Toxic Release Inventory – a program created by the *Superfund* Amendments and Reauthorization Act of 1984 that requires manufacturing facilities and waste handling and disposal sites to report annually on releases of more than 300 toxic materials.

Toxins – poisonous chemicals that react with specific cellular components to kill cells or to alter growth or development in undesirable ways; often harmful, even in dilute concentrations.

Tradeable permits – pollution quotas or variances that can be bought or sold.

Tragedy of the commons – an inexorable process of degradation of communal resources due to selfish self-interest of "free riders" who use or destroy more than their fair share of common property; see *Open access system*.

Transfer station – (waste) a facility allowing drop-off and consolidation of garbage and a wide range of recyclable materials; an integral part of municipal waste management; play an important role in materials recovery and improving transportation economics associated with municipal waste disposal.

Transgenic plant – a plant into which genetic material has been spliced by genetic engineering.

Transitional zone – a zone in which populations from two or more adjacent communities meet and overlap.

Transparent – a substance capable of transmitting light so that objects or images can be seen as clearly as if there were no intervening material.

Transpiration – the process by which water is absorbed by the root system of plants, moves up through the plant, and then evaporates into the atmosphere as water vapor.

Tributary – a small stream that empties into a bigger river.

Triple Bottom Line – a form of sustainability accounting going beyond the financial "bottom line" to consider the social and environmental as well as economic consequences of an organization's activity; generally included with economic accounts; a term coined by John Elkington in 1994.

Trophic level – 1) a step in the movement of energy through an ecosystem; 2) an organism's position in the food chain in an ecosystem.

Tropical – occurring in the tropics (the region on either side of the equator); hot and humid with mean annual temperature greater than 20 °C.

Tropical depression – brings about hurricanes due to change in weather, climate, altitude, latitude, or direction.

Tropical rainforests – forests in which rainfall is abundant (more than 80 inches per year), and temperatures are warm to hot year-round.

Tropical seasonal forest – semi-evergreen or partly deciduous forests tending toward open woodlands and grassy savannas dotted with scattered, drought-resistant tree species; has distinct wet and dry seasons and is hot year-round.

Tropopause – the boundary between the troposphere and the stratosphere.

Troposphere – the layer of air nearest to the Earth's surface; both temperature and pressure usually decrease with increasing altitude; it may be from 4 to 11 miles high (depending on the specific latitude).

Tsunami – giant seismic sea swells that move rapidly from the center of an earthquake; can be 32-65 feet (10-20 meters) high when they reach shorelines hundreds or even thousands of miles from their point of origin.

Tundra – treeless arctic or alpine biome characterized by cold, harsh winters, a short growing season, and potential for frost any month of the year; vegetation includes low-growing perennial plants, mosses, and lichens.

Turbine – a machine for converting the heat energy in steam or high-temperature gas into mechanical energy; in a turbine, the high-velocity flow of steam or gas passes through successive rows of radial blades fastened to a central shaft; see *Tidal energy*.

Turbulence – an eddying motion of the atmosphere that disrupts the normal flow of wind.

Typhoon – a tropical cyclone occurring in the western Pacific or Indian oceans.

U

U-238 – an isotope of uranium (U) used in nuclear power plants.

Ultraviolet (UV) radiation – radiation from the sun that can be useful or potentially harmful; UV rays from one part of the spectrum (UV-A) enhance plant life; UV rays from other parts of the spectrum (UV-B) can cause skin cancer or other tissue damage; the ozone layer in the atmosphere partly shields the Earth from ultraviolet rays reaching the Earth's surface.

Uncontrolled – not under control, discipline or governance.

Unconventional air pollutants – toxic or hazardous substances, such as asbestos, benzene, beryllium, mercury, polychlorinated biphenyls, and vinyl chloride, not listed in the original *Clean Air Act* because they were then thought not to be released in large quantities; see *Non-criteria pollutants*.

Unconventional oil – resources such as *Shale oil* and tar sands that can be liquefied and used as oil.

Undernourished – those who receive less than 90% of the minimum dietary intake over a long-term period; they lack energy for an active, productive life and are more susceptible to infectious diseases.

Underutilized – to utilize less than fully than or below its potential.

Undiscovered resources – speculative or inferred resources or those that haven't yet been thought of.

United Nations – an international organization based in New York and formed to promote international peace, security, and cooperation under a charter signed by 51 founding countries in San Francisco in 1945.

United Nations Framework Convention on Climate Change (UNFCCC) – The UNFCCC and the *Convention on Biological Diversity* (CBD) were established at the 1992 U.N. Conference on Environment and Development in Rio de Janeiro, Brazil; the *Kyoto Protocol* was subsequently formulated by the UNFCCC and sets specific timelines and timetables for reducing industrialized nations' greenhouse gas emissions and allows some international trading of carbon credits.

Universalists – those who believe that some fundamental ethical principles are universal and unchanging; to them, these principles are valid regardless of the context or situation.

Upstream – those processes necessary before an activity is completed (e.g., for a manufactured product, this would be the extraction and transport of materials, etc., that needs to take place before the process of manufacture); see *Downstream*.

Upwelling – the movement of nutrient-rich bottom water to the ocean's surface; can occur far from shore but usually occurs along certain steep coastal areas, where the surface layer of ocean water is pushed away from shore and replaced by cold, nutrient-rich bottom water.

Urban area – an area in which a majority of the people are not directly dependent on natural resource-based occupations.

Urban heat island – the tendency for urban areas to have warmer air temperatures than the surrounding rural landscape, due to the low *Albedo* of streets, sidewalks, parking lots and buildings; these surfaces absorb solar radiation during the day and release it at night, resulting in higher night temperatures; sparse vegetation and paved surfaces increase rain runoff, further reducing cooling effects; temperatures in cities are usually 3-5 °F hotter than the surrounding countryside.

Urban metabolism – the functional flow of materials and energy required by cities.

Urbanization – an increasing concentration of the population in cities and a transformation of land use to an urban pattern of organization.

Useful energy – available energy used to increase system production and efficiency.

Utilitarian conservation – a philosophy that resources should be used for the greatest good for the greatest number for the longest time.

Utilitarianism – a theory that states that the best moral action is also often the most utilitarian one, meaning maximizing the wellbeing of others; see *Utilitarian conservation*.

V

Values – 1) an estimation of the worth of things; 2) a set of ethical beliefs and preferences that determine our sense of right and wrong.

Veloway – cycle track; cycleway; contrasts with the freeway.

Vertical stratification – the vertical distribution of specific subcommunities within a community.

Village – a collection of rural households linked by culture, custom, and association with the land.

Vinyl – a type of plastic (usually PVC) used to make products such as fruit juice bottles, credit cards, pipes, and hoses.

Virtual water – the volume of water required to produce a commodity or service; first coined by Professor J.A. Allan of the University of London in the early 1990s; now more widely known as *Embodied water*.

Viscous – having a relatively high resistance to flow.

Visible light – a portion of the electromagnetic spectrum that includes the wavelengths used for photosynthesis.

Visual waste audit – observing, estimating and recording data on waste streams and practices without physical weighing.

Vitamins – organic molecules essential for life that cannot be manufactured in the body but must be gotten from one's diet; act as enzyme cofactors.

Volatile organic compound (VOC) – molecules containing carbon and differing proportions of other elements such as hydrogen, oxygen, fluorine, and chlorine; with sunlight and heat they form ground-level ozone.

Volt (V) – The unit of electronic force or difference in potential between two points; commonly called voltage; one thousand volts equals 1 kilovolt (kV).

Voluntary simplicity – deliberately choosing to live at a lower level of consumption as a matter of personal and environmental health; see *Simple living*.

Vortex – a spiral motion of fluid within a limited area, especially a whirling mass of water or air that sucks everything near it toward its center.

Vulnerable species – naturally rare organisms or species whose numbers have been so reduced by human activities that they are susceptible to actions that could push them into a threatened or endangered status.

W

Warm front – a long, wedge-shaped boundary caused when a warmer advancing air mass slides over neighboring cooler air parcels.

Waste – any material (liquid, solid or gaseous) that is produced by domestic households or commercial, institutional, municipal or industrial organizations and which cannot be collected and recycled in any way for further use; for solid wastes, this involves materials that currently go to landfills, even though some of the material is potentially recyclable.

Waste analysis – the quantifying of different waste streams, the recording and detailing of it as a proportion of the total waste stream, determining its destination, and recording details of waste practices.

Waste assessment/audit – observing, measuring, and recording data and collecting and analyzing waste samples. Some practitioners consider an assessment to be one where observations are carried out visually, without sorting and measuring individual streams; see *Visual waste audit*.

Waste avoidance – a primary pillar of the waste hierarchy; avoidance works on the principle that the greatest gains result from efficiency-centered actions that remove or reduce the need to consume materials in the first place but still deliver the same outcome.

Waste factors – (used in round-wood calculations) give the ratio of one cubic meter of round wood used per cubic meter (or ton) of the product made from it.

Waste generation – generation of unwanted materials, including recyclables as well as garbage; waste generation = materials recycled + waste to landfill.

Waste hierarchy (waste management hierarchy) – a concept promoting waste avoidance ahead of recycling and disposal, often referred to in community education campaigns as "reduce, reuse, recycle;" recognized in the Environment Protection Act of 1970, promoting management of wastes in this order of preference: avoidance, reuse, recycling, recovery of energy, treatment, containment, disposal.

Waste lagoons – a blocked-off area, used for the dumping of waste products.

Waste management – practices and procedures that relate to how the waste is dealt with.

Waste minimization – 1) techniques to keep waste generation at a minimum level to divert materials from landfill and thereby reduce the requirement for waste collection, handling, and disposal to a landfill; 2) recycling and other efforts made to reduce the amount of waste going into the waste stream.

Waste reduction – measures to reduce the amount of waste generated by an individual, household or organization.

Waste stream – Waste materials that are either of a particular type (e.g., "timber waste stream") or produced by a particular source (e.g., "C & I waste stream"); the steady flow of varied wastes, from domestic garbage and yard wastes to industrial, commercial and construction refuse.

Waste treatment – where some additional processing is undertaken of a waste; may be done to reduce its toxicity or increase its degradability or compostability.

Wastewater – used water; generally, not suitable for drinking.

Water consumption – in water accounting, distributed water use + self-extracted water use + reuse water use - distributed water supplied to other users - in-stream use (where applicable).

Water cycle – the recycling and utilization of water on Earth, including atmospheric, surface and underground phases and biological and non-biological components.

Water droplet coalescence – a mechanism of condensation that occurs in clouds too warm for ice crystal formation.

Water entitlement – the entitlement, as defined in a statutory water plan, to a share of water from a water source.

Water footprint – the total volume of freshwater that is required in a given period to perform a particular task or to produce the goods and services consumed at any level of the action hierarchy; a country's water footprint is a concept introduced by Hoekstra in 2002 as a consumption-based indicator of overall water use in a particular country — the volume of water needed to produce all the goods and services consumed by the inhabitants of a country.

Water harvesting – see *Rainwater harvesting*.

Water intensity – the volume of water used per unit of production or service delivery; this is generally further reduced to monetary unit return per given volume of water used; essentially equivalent to *Water productivity*.

Water productivity (WP) – the efficiency of outcomes for the amount of water used; the quantity of water required to produce a given outcome; WP-field relates to crop output (e.g., kg of wheat produced per m^3 of water); WP-basin relates to water productivity in the widest possible sense, including crop, fishery yield, environmental services, etc.; increasing WP means obtaining increasing value from the available water.

Water quality – the microbiological, biological, physical and chemical characteristics of water.

Water resources – water in various forms, such as groundwater, surface water, snow, and ice, at present in the land phase of the hydrological cycle — some parts may be renewable seasonally, but others may be effectively mined.

Water restrictions – mandatory restrictions on the use of water, imposed relative to water storage levels.

Water stress – a situation when residents of a country or region do not have accessible enough, high-quality water to meet their everyday needs.

Water table – the surface between the zone of saturation and the zone of aeration; water seeping down from rain-soaked surfaces will sink until it reaches an impermeable or water-tight layer of rock; the water will collect above this layer, filling all the pores and cracks of the permeable portions; the top of this area of water is called the water table.

Water trading – transactions involving water access entitlements or water allocations assigned to water access entitlements.

Water treatment – the process of converting raw, untreated water to a public water supply safe for human consumption; can involve, variously, screening, initial disinfection, clarification, filtration, pH correction, and final disinfection.

Waterlogging – water saturation of soil that fills all air spaces and causes plant roots to die from lack of oxygen; often the result of over-irrigation.

Watershed – the area of land that catches rain and snow and drains or seeps into a marsh, stream, river, lake or groundwater; often contained in the area of land between two ridges of high land, which divide two areas that are drained by different river systems.

Watt, Kilowatt – a unit of measure of electric power at a point in time as capacity or demand; 1 watt = 1 joule/second; 1 joule = energy spent in one second when a current of 1 amp flows through a resistance of 1 ohm; 1 kilowatt – 1000 watts.

Weather – the hourly/daily change in atmospheric conditions (moisture, temperature, pressure, and wind), which over a longer period constitute the climate of a region; see *Climate*.

Weathering – changes in rocks brought about by exposure to air, water, changing temperatures and reactive chemical agents.

Well-being – a context-dependent physical and mental condition determined by the presence of basic material for a good life, freedom and choice, health, good social relations, and security.

Wetlands – areas of permanent or intermittent inundation, whether natural or artificial, with water that is static or flowing, fresh, brackish or salt, including areas of marine water not exceeding 6 meters (19 feet) at low tide; engineered wetlands are becoming more frequent and are sometimes called *constructed wetlands*; in urban areas wetlands are sometimes referred to as the kidney of a city.

Whitegoods – household electrical appliances like refrigerators, washing machines, clothes dryers, and dishwashers.

Wicked problems – problems with no simple right or wrong answer where there is no single, generally agreed-on definition of or solution for the particular issue.

Wilderness – an area of undeveloped land affected primarily by the forces of nature; an area where humans are visitors who do not remain.

Wilderness Act – legislation of 1964 recognizing that leaving the land in its natural state may be the highest and best use of some areas.

Wildlife – plants, animals, and microbes that live independently of humans; plants, animals, and microbes that are not domesticated.

Wildlife refuges – areas set aside to shelter, feed, and protect wildlife; however, due to political and economic pressures, refuges often allow hunting, trapping, mineral exploitation, and other activities that threaten wildlife.

Wind – moving air, especially a natural and perceptible movement of air parallel to or along the ground.

Wind energy – the kinetic energy present in the motion of the wind; can be converted to mechanical or electrical energy; a traditional mechanical windmill can be used for pumping water or grinding grain; a modern electrical *Wind turbine* converts the force of the wind to electrical energy for consumption on-site and/or export to the electricity grid.

Wind farms – large numbers of windmills concentrated in a single area; usually owned by a utility or large-scale energy producer.

Wind turbines – a turbine with a large fanned wheel that spins in the wind to generate electricity; see *Wind energy*.

Windbreak – rows of trees or shrubs planted to block wind flow, reduce soil erosion, and protect sensitive crops from high winds.

Wise Use Groups – a coalition of ranchers, loggers, miners, industrialists, hunters, off-road vehicle users, land developers, and others who call for unrestricted access to natural resources and public lands.

Withdrawal – a description of the total amount of water taken from a lake, river or aquifer.

Woodland – a forest where tree crowns cover less than 20 percent of the ground; see *Open canopy*.

Work – 1) physical or mental effort; 2) a force exerted for a distance; 3) an energy transformation process which results in a change of concentration or form of energy.

World conservation strategy – a proposal for maintaining essential ecological processes, preserving genetic diversity and ensuring that utilization of species and ecosystems is sustainable.

World Trade Organization (WTO) – an association of 135 nations that meet to regulate international trade.

X

Xeriscaping – landscaping with drought-resistant plants that need no watering.

X-ray – very short wavelength in the electromagnetic spectrum; can penetrate soft tissue; although it is useful in medical diagnosis, it also damages tissue and causes mutations.

Y

Yellowcake – the concentrate of 70 to 90% uranium oxide extracted from crushed ore.

Yucca Mountain, Nevada – the U.S. Department of Energy's potential underground geological repository for spent nuclear fuel and high-level radioactive waste.

Z

Zebra mussel – a European and Asian freshwater mussel regarded as a nuisance in the Great Lakes and surrounding waterways, where it was accidentally introduced; Latin name *Dreissena polymorpha*.

Zero population growth (ZPG) – the number of births at which people are just replacing themselves; also called the "replacement level of fertility."

Zero waste – turning waste into a resource; the redesign of resource-use so that waste can ultimately be reduced to zero; ensuring that by-products get used elsewhere, and goods are recycled, in emulation of the cycling of wastes in nature.

Zone of aeration – zone immediately below the ground surface within which pore spaces are partly filled with water and partly filled with air.

Zone of leaching – the layer of soil just beneath the topsoil where water percolates, removing soluble nutrients that accumulate in the subsoil.

Zone of saturation – lower levels of soil where all spaces are filled with water.

Glossary of Terms

Figure 1.1 (a, b). *A geological time scale constructed by The Geological Society of America.*
http://www.geosociety.org/science/timescale/

Figure 1. 2. *Convergent and divergent plate boundaries that occur in oceans and land.*
http://oceanexplorer.noaa.gov/facts/plate-boundaries.html

Figure 1.3. *A simplified map of the tectonic plates that span the Earth.*
https://volcanoes.usgs.gov/about/edu/dynamicplanet/ballglobe/simplifiedmap.pdf

Figure 1.4. *A seismograph and spring for the motion of the Earth.*
http://earthquake.usgs.gov/learn/glossary/?termID=167

Figure 1.5. *Triangulation method to determine the epicenter of an earthquake.*
http://earthquake.usgs.gov/learn/kids/eqscience.php

Figure 1.6. *Equinoxes and solstices and their time frames.*
http://www.weather.gov/cle/Seasons

Figure 1.7. *Seasons are a product of the Earth's tilted axis and its path around the Sun.*
http://spaceplace.nasa.gov/seasons/en/

Figure 1.9. *The solar radiation received by the surface of the Earth.*
http://earthobservatory.nasa.gov/Features/EnergyBalance/page2.php

Figure 1.10. *Aurora in Alaska.*
http://www.nasa.gov/content/goddard/purple-and-green-aurora-in-alaska

Figure 1.11. *Layers of the atmosphere and several phenomena that occur across it.*
http://www.srh.noaa.gov/jetstream/atmos/layers.htm

Figure 1.13. *Comparison of water temperatures in three El Niño events.*
http://www.weather.gov/lot/El_Nino

Figure 1.14. *The division of fresh and saltwater sources on Earth.*
http://water.usgs.gov/edu/earthwherewater.html

Figure 1.15. *The flow pattern of the five major ocean-wide gyres.*
http://oceanservice.noaa.gov/education/kits/currents/05currents3.html

Figure 1.16. *Spray system that conserves water.*
http://water.usgs.gov/edu/irsprayhigh.html

Figure 1.17. *Drip/micro irrigation methods using horizontal pipes.*
http://water.usgs.gov/edu/irdrip.html

Figure 2.1. *Picture of the Earth from space.*
NASA: Mission AS17. 1972. Apollo 17 Crew.

Figure 2.2. *Dry creek bed at Quivira National Wildlife Refuge.*
Laubhan, Rachel. 2012.

Figure 2.3. *A Hutchinsonian niche: purple-throated Carib's bill.*
Sharp Photography. *Morne Diablotins National Park in Dominica.*

Figure 2.4. *Antagonism: the black walnut tree's roots.*
L., Juglans Nigra. *Black Walnut.*

Figure 2.5. *The jaguar is an example of a keystone species.*
Foresman, Pearson Scott. *Line Art Drawing of a Jaguar.*

Figure 2.6. *Farles Prairie in Ocala National Forest.*
Sandra Fried. USDA. Forest Service.

Figure 2.8. *Freshwater aquatic and terrestrial food web.*
Thompson, Mark David. "A Freshwater Aquatic and Terrestrial Food Web."

Figure 2.9. *Ecological energy flow.*
Creative Commons, 2015.

Figure 2.10. *North Sea phytoplankton bloom.*
Jesse Allen. Oregon State Univ., Jochen Wollschläger, Helmholtz-Zentrum Geesthacht.

Figure 2.11. *Charles Darwin, English naturalist and geologist.*
Edwards, Ernest. 1867. *Charles Darwin.*

Figure 2.12. *Shells and fossils.*
Daguerre, Louis. *Daguerreotype of Louis Daguerre Ca. 1850. Kiperpipa.*

Figure 2.13. *Satellite photograph of clouds caused by the exhaust from ship smokestacks.*
Liam Gumley. Space Science and Engineering Center, U. of Wisconsin-Madison.

Figure 2.14. *Succession after a wildfire in a boreal pine forest.*
Hannu. Lahemaa National Park, Estonia.

Figure 2.15. *Water cycle.*
John M. Evans. 2005. U.S. Geological Survey. U.S. Department of the Interior.

Figure 2.16. *Phosphorus cycle.*
United States. Environmental Protection Agency.

Figure 2.17. *Sulfur formations in White Island, New Zealand.*
Various Sulfur Formations on White Island. 2014.

Figure 2.18. *Mikhail Lomonosov, Russian scientist. The law of mass conservation, 1756.*
Miropolskiy, Leontiy & Prenner. 1787. Russian Academy of Sciences, St. Petersburg.

Figure 3.1. *This world map indicates the average population density per km².*
Miguel Contreras. 2006.

Figure 3.2. *Line graph plots human exponential population growth between 1800 and 2000.*
Clevercapybara. 2015.

Figure 3.4. *This life table, for mortality levels and trends for children under 5.*
Simon Berry. 2013. Flickr

Figure 3.6. *London, the most populous city in the world at the turn of the 20th century.*
John L. Stoddard. 1900.

Figure 3.7. *United States map indicates the 2013 total fertility rates by state.*
Ali Zifan. 2015.

Figure 3.9. *ECOSOC world map highlighting the current LEDCs.*
Gabbe. 2015.

Figure 3.10. *This government sign in Nanchang for birth control.*
Venus. 2006.

Figure 3.11. *Planetary boundary diagram.*
Ninjatacoshell. 2015.

Figure 3.12. *Diagram showing the Ebola animal-animal and animal-human transmission cycle.*
Centers for Disease Control and Prevention (CDC).

Figure 3.13. *Thirty-two children in a small classroom near Nagar, Pakistan.*
Shaun D. Metcalfe. 2009.

Figure 3.14. *World population from U.S. Census Bureau overlaid with fossil fuel use.*
Vaclav Smil. *Energy Transitions: History, Requirements, Prospects.*

Figure 3.15. *Correlation between human population and species lost to extinction.*
Scott, J.M. 2008. Idaho Cooperative Fish and Wildlife, University of Idaho.

Figure 4.1. *Cranberry harvest, New Jersey*
Keith Weller, Agricultural Research Service, USDA.

Figure 4.2. *Deforestation of peat forest in Indragiri Hulu, Sumatra, Indonesia.*
Riau deforestation. Wakx. 2007.

Figure 4.3. *Pesticide application, Yuma, Arizona.*
Jeff Vanuga. 2011. Nation Resources Conservation Service, USDA.

Figure 4.4. *DDT solution sprayed onto sheep for tick control in Benton County, Oregon, 1948.*
Robert W. Every, 1948. United States Office of Special Counsel.

Figure 4.5. *A forest on San Juan Island, Washington.*
Tom Harpel. 2004. Forest on San Juan Island.

Figure 4.6. *A yellow birch in the Allegheny National Forest, Pennsylvania*
Nicholas A. Tonelli. 2011. Old-growth forest.

Figure 4.7. *The Brins Fire, Sedona, Arizona, 2006.*
Coffeegirlyme. 2006. Sedona fire.

Figure 4.8. *Rangeland of the Red Desert, Wyoming.*
Sam Cox, USDA. 2009. Muddy Water Red Desert.

Figure 4.9. *Dried out soil, Sonora, Mexico.*
Tomas Castelazo. 2007.

Figure 4.10. *Major uses of land, 1959-2007.*
www.ers.usda.gov. USDA, Economic Research Service, Major Land Use data product.

Figure 4.11. *Suburban sprawl, Rio Rancho, New Mexico.*
Rio Rancho sprawl. Riverrat. 2012.

Figure 4.12. *USS America aircraft carrier passes through the Suez Canal, Egypt, 1981.*
W. M. Welch. 1981. USS America (CV-66). U.S. Navy.

Figure 4.13. *Mount Saint Elias, Wrangell-St. Elias National Park and Preserve, Alaska.*
David Sinson. 2005. National Oceanic and Atmospheric Administration. Mt. Saint Elias.

Figure 4.14. *Deepwater Horizon oil spill, Gulf of Mexico, 2010.*
Justin Stumberg. 2010. United States Navy.

Figure 4.15. *U.S. Mining Fatalities CY 1978-2014. 2016.*
http://arlweb.msha.gov/MSHAINFO/FactSheets/MSHAFCT10.asp

Figure 4.16. *Open-pit mining, Sunrise Dam Gold Mine, Australia.*
Callistemon. 2010.

Figure 4.17. *The Farmington Mine disaster, West Virginia.*
Mine Safety and Health Administration, U.S. Department of Labor. 1968.

Figure 4.18. *Alaskan fishermen, 1927.*
W.B. Miller. 1927. U.S. Library of Congress.

Figure 4.26. *Aquaculture in Luoyuan Bay, Fuzhou, China.*
Jack Parkinson. 2006.

Figure 4.27. *Global total fish harvest.*
https://commons.wikimedia.org/wiki/File:Global_total_fish_harvest.svg

Figure 4.28. *Gulf Coast wetland draining, and flooding by Hurricane Katrina.*
National Oceanic and Atmospheric Administration. 2005.

Figure 4.29. *John Day Dam and fish ladder, Columbia River.*
United States Army Corps of Engineers. 2005.

Figure 5.1. *The exponential growth of the human population throughout history.*
http://www.esrl.noaa.gov/gmd/infodata/lesson_plans/Connecting%20Population%20and
%20%20Climate%20Change.pdf

Figure 5.2. *World total final consumption from 1971 to 2012 by fuel.*
http://www.iea.org/publications/freepublications/publication/KeyWorld2014.pdf

Figure 5.3. *U.S. primary energy consumption estimates by source, 1950 to 2011.*
http://www.eia.gov/todayinenergy/detail.cfm?id=9210

Figure 5.4. *The total world energy consumption by the source in 2013.*
www.ren21.net/Portals/0/documents/Resources/GSR/2014/GSR2014_full%20report_low
%20res.pdf

Figure 5.5. *Primary energy use by the source in the United States in 2013*
http://www.eia.gov/totalenergy/data/monthly/archive/00351405.pdf

Figure 5.6. *Global consumption of petroleum by country in 2013.*
http://www.eia.gov/beta/international/

Figure 5.7. *World Oil Reserves, 2013.*
https://commons.wikimedia.org/wiki/File:Oil_Reserves.png

Figure 5.8. *U.S. Oil production and imports (in millions of barrels per day).*
http://www.energy.gov/science-innovation/energy-sources/fossil

Figure 5.9. *Recoverable Coal Reserves by Country, 2011.*
http://www.eia.gov/cfapps/ipdbproject/IEDIndex3.cfm?tid=1&pid=1&aid=2

Figure 5.10. *Nuclear fuel cycle; uranium recovery, enrichment, utilization, and disposal.*
nrc.gov, public domain

Figure 5.11. *Typical hydroelectric dam and the route which water takes to generate*
*electricity. http://*water.usgs.gov/edu/hyhowworks.html

Figure 5.12. *A "Zero Energy Home" located in New Paltz, New York.*
http://energy.gov/eere/buildings/doe-tour-zero-preserve-greenhill-contracting

Figure 5.14. *Emissions & water savings using EPA's AVERT.*
http://www.energy.gov/eere/articles/earth-friendly-wind-vision

Figure 5.15. *An example of a small hydropower system on personal property.*
http://energy.gov/energysaver/microhydropower-systems

Figure 5.16. *The Agucadoura Wave Farm off the shore of Portugal*
www.boem.gov/Renewable-Energy-Program/Renewable-Energy-Guide/Ocean-Wave-Energy.aspx

Figure 6.1. *Exhaust from a diesel truck.*
The United States Environmental Protection Agency (EPA). 2008.

Figure 6.4. *Acid deposition.*
The United States Environmental Protection Agency (EPA). 2009.

Figure 6.5. *Urban heat island profile.*
United States National Oceanic and Atmospheric Administration. 2008.

Figure 6.6. *Chamber for measuring volatile organic compounds emitted from furnishings.*
Tracey Nicholls. Australia's Commonwealth Scientific & Industrial Research
Organization. CSIRO.

Figure 6.7. *President Lyndon B. Johnson signed the Clean Air Act in 1967.*
Mike Geissinger. 1967. Executive Office of the President of the U.S. *LBJ Library.*

Figure 6.8. *Nutrient pollution: runoff of fertilizer during heavy rain.*
Lynn Betts. 1999. USDA, Natural Resources Conservation Service.

Figure 6.9. *Marine debris on a Hawaii beach.*
U.S. National Oceanic and Atmospheric Administration. 2008. *Marine Debris Program.*

Figure 6.11. *Groundwater pollution: a pit latrine in Lusaka, Zambia, pollutes the nearby well.*
Mayumbelo, Kennedy. 2009. Sustainable Sanitation Alliance.

Figure 6.13. *Hazardous waste collection site.*
Robert Kaufmann. 2006. *FEMA.* U.S. Department of Homeland Security.

Figure 6.14. *Mercury waste from the Brunswick Pulp and Paper Company*.
Paul Conklin. 1973. National Archives and Records Administration: Still Picture Records Section.

Figure 6.15. *Workers test a leachate collection system at the Savannah River Site.*
The United States Department of Energy. 2012.

Figure 6.16. *The United States Environmental Protection Agency logo.*
The United States Environmental Protection Agency.

Figure 6.17. *A worker labels PCB-containing transformer.*
The United States Army Corps of Engineers.

Figure 6.18. *Fluor Fernald workers pneumatically remove hazardous waste.*
The United States Department of Energy. 2005.

Figure 6.19. *Deep injection well for disposal of hazardous, industrial and municipal wastewater.*
The United States Environmental Protection Agency. 2010.

Figure 6.20. *Debris and injured birds caused by water pollution.*
Leary Pete. 2013. The United States Fish and Wildlife Service.

Figure 6.21. *Demand curve showing the microeconomic concept of a negative externality.*
Bandersnatch, Struthious. 2011.

Figure 6.22. *Smog-damaged plant at the statewide air pollution research center.*
Gene Daniels. 1972. National Archives and Records Administration: Still Pictures Records Section.

Figure 6.23. *Sustainable development.*
Dréo, Johann. 2006.

Figure 7.1. *Stratospheric ozone production.*
http://www.esrl.noaa.gov/csd/assessments/ozone/2010/twentyquestions/Q2.pdf

Figure 7.2. *Various types of radiation that compose the electromagnetic spectrum.*
http://www3.epa.gov/climatechange/kids/basics/today/greenhouse-effect.html

Figure 7.3. *Ultraviolet (UV) photons harm the DNA molecules of living organisms in various ways.* http://earthobservatory.nasa.gov/Features/UVB/

Figure 7.4. *A condensed cycle outlining the greenhouse effect.*
http://www3.epa.gov/climatechange/kids/basics/today/greenhouse-effect.html

Figure 7.5. *Worldwide greenhouse gas emissions by economic sector.*
https://www3.epa.gov/climatechange/ghgemissions/global.html

Figure 7.6. *Fluorinated gas emissions in the United States, by source:1990-2013.*
http://www3.epa.gov/climatechange/ghgemissions/gases/fgases.html

Figure 7.7. *Nitrous Oxide emissions in the United States by source. 1990-2013.*
http://www3.epa.gov/climatechange/ghgemissions/gases/n2o.html

Figure 7.8. *The United States carbon dioxide emissions by source. 1990-2013.*
http://www3.epa.gov/climatechange/ghgemissions/gases/co2.html.

Figure 7.9. *U.S. methane emissions; Greenhouse Gas Emissions and Sinks: 1990-2013.*
http://www3.epa.gov/climatechange/ghgemissions/gases/ch4.html

Figure 7.10. *A polar bear and the effects of climate change on habitat loss.*
http://www.fws.gov/alaska/fisheries/mmm/polarbear/esa.htm

Figure 7.11. *The process of coral bleaching.*
http://oceanservice.noaa.gov/facts/coral_bleach.html

Figure 7.12. *Adult alfalfa weevils threatened the native species through foraging*
http://www.agriculture.gov.sk.ca/Alfalfa-Weevil

Figure 7.13. *Species that are endangered or threatened as of November 28, 2015.*
U.S. Fish and Wildlife Service

Figure 7.14. *A wildlife crossing over a highway that allows for the safe crossing of animals.*
http://www.fhwa.dot.gov/environment/critter_crossings/tortoise.cfm

Made in the USA
Monee, IL
17 November 2022

17995148R00256